THE MOUNTAIN

The Weisshorn from above Täsch Alp.

THE MOUNTAIN

*Renewed Studies in Impressions
and Appearances*

by
John C. Van Dyke

Foreword
by
Peter Wild

University of Utah Press
Salt Lake City

Library of Congress Cataloging-in-Publication Data

Van Dyke, John Charles, 1856–1932.
    The mountain : renewed studies in impressions
and appearances / by John C. Van Dyke ; foreword
by Peter Wild.
        p.   cm.
    Originally published: New York : C. Scribner's
Sons, 1916.
    ISBN 0–87480–387–X (alk. paper)
    1. Mountains   I. Title.
GB511.V3   1992
551.4'32—dc20   91–51096
                                                       CIP

CONTENTS

FOREWORD

About a hundred and twenty years ago, a band of
Sioux Indians left its home camp in eastern Minne-
sota and struck out into the unsettled western plains.
This was the Indians' traditional summer hunt to fat-
ten their larders with buffalo meat for the lean winter
months to come. Nothing unusual marked the trip,
except one thing. With the group rode a white boy
in his early teens. He was their young friend, John
C. Van Dyke, the son of a wealthy family recently
moved from crowded New Jersey to the western
frontier.

The future art critic was entering a world far wilder
than his pioneer homestead on the wooded banks
of the Mississippi River, and the experiences ahead—
this brief gaze into the disappearing primitive beauty
of the continent—would change him forever. Over
the following weeks and months the little group
roamed, like a ship on an untraveled sea, for thousands of miles, ever farther into what was then a
mostly unsettled kingdom. Out there, remembers
Van Dyke, "All space was ours, time was not...."
Following the herds, killing buffalo as they went, the
Indians crossed the Bad Lands of present-day South

Dakota and zigzagged west until the Big Horn Mountains of the Wyoming Territory loomed before them. They swam the Yellowstone River and somewhere in central Montana turned north to begin a leisurely circle homeward. Weeks later, they once again recognized what Van Dyke acidly calls "the savage edge of civilization."

In white men's terms, it was a strenuous, if not heroic, trip. Young Van Dyke rode a half-wild Indian pony, bareback or with a moose hide for a saddle; with the rest of his friends, the boy scrounged for food and slept on the ground. With the band he dashed up and down ravines in bone-jolting pursuits and endured the hot plains sun. Broken bones, snakebite, and attacks by hostile Cheyennes were simply what one risked on such adventures, but risked stoically: "To look out for yourself, to ask no questions, and to make no murmur" were the Indians' rules of the road. There were no whiners among these hunters.

Chances are, however, that neither heroism nor hardship were foremost in the boy's mind. At the time, John probably had little idea how unusual his trip was in the historical sense. And as to hardships, the young pioneer already had learned to accept them as a matter of course, learned that a person either rose above difficulties or let them drag him down. Van Dyke concentrated on the important things, the immediate and continuing thrill of the trip, the "mad fascination" of exploring new lands. Not only that but reveled in the visual treasures all around him, the "clouds of ducks pouring in at dusk" over a prairie lake, the delicate lilac hue of his horse's moving shadow. Years later he recalls the scouts sent out from camp to make a final check for enemies before night fell, the "half-naked Sioux silhouetted against the blood-red twilight, each one bunched over his pony's shoulder and peering catlike into the gathering gloom."

These are not likely the details most of us would remember decades later or express so sharply, with such poetry and precision. There's a reason why Van Dyke did. He was blessed with a painter's eye and the same talent for seeing that later made him the nation's foremost Rembrandt scholar. The same passion for nature's forms and colors that turned him into an avid devotee of the Art for Art's Sake movement also shaped his boyhood perceptions of ducks descending and the changing colors of his horse's shadow gliding alongside him over the plain's deep grass.

Yet what impressed Van Dyke indelibly on the trip was his first glimpse of the Rocky Mountains. Surrounded by his stern companions, he sat astride his pony gazing at the "blue-and-silver wonder" of the distant peaks. "For days they hung upon our horizon," says Van Dyke of the vast specter. "We looked to them in the morning and saw them glittering white, at noon we saw them grow pale and shadowy under the midday heat, and at night the last shaft of the sunset fire was reflected rose-hued from their summits." Looking back on the event, Van Dyke knew that his friends were not so easily moved, for they were practical men, "seeking buffaloes, not mountains." But the romantic in him knew that he was changed by the presence, by "the most lasting impression of all" in what would be a lifetime of visual feasts.

Fittingly for the passionate romantic yearning for the unattainable, the group then swung north and east. Van Dyke didn't reach the Rockies after his first view. But his mind's eye continued dazzled, and though he later would become intimately familiar with the range, that first glimpse—as of Dante looking upon Beatrice—became his aesthetic paragon. Though in the coming pages Van Dyke introduces the reader to the beauty of mountains in Switzerland, Russia, Mexico, Canada, Spain—to the beauty of mountains around the world—his standard of comparison remained the Rockies. By titling his book in the singular rather than the plural, he hews to the remembered ideal he carries in his head.

Yet why did Van Dyke choose mountains and not some other landform for his praises? As he says, mountains do offer an "extraordinary display" of forms, and they do remind us of the earth's elemental forces. Yet *The Mountain* needs to be placed in context. In his travels around the globe, Van Dyke saw all aspects of nature as good and beautiful in themselves, and in fact he wrote several volumes celebrating by turns America's deserts, the world's seas, the Grand Canyon, and the green and gentle landscapes around New Brunswick, New Jersey, where he spent a lifetime teaching art history at Rutgers College (now University). In each book Van Dyke treats his subject as if at last he's discovered the ideal of aesthetic delights. His desert study exults: "And there you have the most decorative landscape in the world, a landscape all color, a dream landscape." Years later and thousands of miles away, Professor Van Dyke stands on Italy's Alban Hills looking down as sunset fades and Rome's lights twinkle on. To him, the scene is "the very poetry of vision." Such declarations say far more about Van Dyke's passionate enthusiasms for the topic at hand than they call into question the sincerity of a person who saw sensuously wherever he traveled.

One wonders, too, how Van Dyke manages to bring it off, how he manages to write whole books about the aesthetic pleasures of viewing oceans or mountains without boring his readers with artistic theory or leaving them behind on his airy flights. Despite his surging emotions, Van Dyke is a calculating writer and a consummate organizer. In *The Mountain* he gives a framework to his passions with chapters devoted to foothills, the peaks above timberline, glaciers, and such. He reserves entire sections for comparing and contrasting the visual impressions of the world's mountain lakes and for musing on the particular quality of mountain skies. Yet no *Luftmensch*, Van Dyke grounds his writing in solid science, explaining, for instance, the geology of mountain building, even, far ahead of his times, hinting at a theory of tectonic plates. He discusses life zones and the physics of various lighting effects. The science lends his writing credibility.

But it also is something of a ruse. For in the final analysis, the factual aspects of nature, interesting as they may be, are not what fascinates Van Dyke. He reminds us that "science eventually grows weary of its own hypotheses"; theories change, and, in any case, the facts of science are but the accidental bases for what delights our eyes. Ultimately, Van Dyke is a sensualist. He may tell us about lateral moraines, frontal moraines, and eskers—and do so clearly, with authority—but what sets his books apart are their passionate descriptions, of a night sky so deep and dark that the stars blaze to gold, of the peculiar sounds quivering through a Swiss glacier like "threads of pain or cries of souls in bondage." The achievement of Van Dyke's books is not so much that they inform us as that they change us, teaching us to see and hear anew and in ways that ever after enrich our own experiences in nature.

This book reprints the original edition of *The Mountain*, published by Scribner's in 1916. The occasional notes are Van Dyke's. Ah, but one last thing. I said a bit earlier that Van Dyke is a calculating writer. He also can be a mischievous one. *The Mountain* is an emotionally heady but serious work. Still, even here, the professor is having a little fun. As will be seen, the volume contains snatches of poetry at times not identified by Van Dyke, though some of it obviously is by Coleridge, Byron, Shelley, and other well-knowns. In places the verse supports a point made in prose, but readers need to be on the watch. For in other cases this is but the apparent purpose, as is the one-line quote from Byron on page 178. In actuality, Van Dyke is setting up his readers. On page 223 his larger end comes clear; it is to show how even the Greats often fall short in describing a nature glorious beyond the capabilities of the human arts. That some of the poetry is Van Dyke's own, presented in company with lines by Shelley and Goethe, only heightens the humor. On the one hand, Van Dyke—by his own admission elsewhere not more than an adequate poet—jokes that his verses can stand head and shoulders with those of renowned poets. On the other hand, he's making a comment going the other way, that his verses, too, pale by comparison to nature's supreme effects. Such is but one of the complexities gone unrecognized in Van Dyke over the years and only now being discovered through renewed attention to his writing.

TUCSON, ARIZONA   PETER WILD

PREFACE–DEDICATION

𝕿𝖔

B. R. C.

OUR first fancy in mountain scenery is usually the spectacular, the startling, perhaps the fantastic. The spurs and spines of the Dolomites, the sharp shaft of the Matterhorn, the beetling height of Kinchinjunga command our wonder by their extraordinary display of form. Years after, when the ranges have become a more familiar story, we perhaps care less for the abrupt than the quiescent and are content with the flowing lines of the Appalachians or the rounded forms of the Scotch hills.

We change our point of view, but the mountains themselves change not. Whatever face they turn to us, whatever their appearance, it is always one of beauty. They are mere breaks in the crust—crushings of the earthshell—and by the same reckoning color is a mere break in the beam of light, but what a wonder work each of them spreads before us! We pick and choose here and there—some for the strength of form and others for the glory of color—but nature herself makes no choice and has no preference. For she formed and garbed and hued each manifestation, from the first gentle heave of a foot-hill to the last sun-flash on a pinnacled peak, with a singular care and endowed it with a peculiar and a superlative beauty. Each after its kind was made perfect and complete within itself.

When nature sets such a pace what better can one do than follow? In these studies of mountains I have not given either preference or illustration to any one form of mountain beauty. I have spoken much of the Rockies and devoted the first chapter to the long approach to them across the prairies of Dacotah; but I have also taken many illustrations from the Alps, the Carpathians, the Caucasus, the Apennines, the Mexican Sierras—some of them quite as familiar to Americans as their own mountains. From all sources and from many lands I have tried to show the varied features and different types of mountain splendor. The variety is infinite, and the Ruskins who have formed theories of mountain form and color from the Alps alone fall short in their generalizations. There is a peculiarity—one might say a mountain individuality—about not only the peaks but the ranges. No two of them are alike.

For the rest these sketches should speak for themselves. They will tell you that my interest is with the beauty of the mountains rather than their science, their history, or their conquest by Alpinists. I have used science herein by way of illustrating the bases of form and color and used history and narrative merely as a setting for the picture. It is the picture—the pictorial in the mountains—that I am seeking, and the remainder is only so much pigment, varnish, and gilt frame. All the world loves the mountains—loves them as color patterns on the blue, as marvels of light and shade, as symbols of peace and purity—loves them, in other words, as one might a picture. That there is a supreme grandeur about the peaks and a great beauty in the lesser hills every one will acknowledge.

But the grandeur of the peaks is as hard to capture with the brush as the beauty of the hills with the pen. The landscape-painter and the writer have had very moderate success in portraying them. The great features of nature seem defiant of representation. Whether this attempt of mine is better or worse than others I know not. It may have a saving grace of difference. For the greater part of it is record of my own seeing carried over a long number of years. I shall have to insist that the record is true so far as it goes—that is, true to my point of view.

J. C. V. D.

San Bernardino Mountains
August, 1915.

THE MOUNTAIN

CHAPTER I

FROM AFAR

Dacotah! The Dacotah of the Sioux! Before the railways crossed it, before a wagontrail was broken, before ever the white man came, what a wilderness it was! The prairie-grass reached up to the pony's knees, the sun came up out of it and went down into it, a wide horizon stretched around one in a ring, a light-shot sky of blue came down to meet it. The buffalo, the gray wolf, and the coyote were its occupants. It was Any Man's Land, it was No Man's Land. The antelope, with restless, bulging eyes, kept watch about him in a circle. And the human who ventured there, whether white man or red, never relaxed his vigilance. The right of might was the law of the prairie, and from it there was no appeal. Life was for those who could keep it.

The great prairie.

Rare, indeed, was the sight of man in that wilderness. There were few camps and no settlements. Even the Indians, seen occasionally in small, bead-like strings crossing the wide expanse, seemed somewhat out of place, somewhat like shore birds lost at sea, winging a weary way over endless swales and divides. They were never too fond of the grass prairies, for they lacked in protection. They crossed and recrossed them on hunting expeditions, they camped on them, hunted on them; but they never cared to live on them. Almost always the band was small and moving toward the west—toward the northern range in Montana, where the buffalo herds were plentiful.

The way led across the Missouri, through the Bad Lands, over into the valleys of the Powder, the Tongue, and the Big Horn. It was a wild, wild way, but what a lure it had for the venturesome! There is always a mad fascination about a new and unexplored country.

It required something of enthusiasm and more of endurance to ride with the Sioux in those days. Your horse was only a mustang, half bridle-wise and wholly wicked, that rolled you under in swimming a river and pounded you with his fore feet when he got his head; your stirrups, into which you thrust moccasined feet, were mere loops in a buckskin thong flung over the pony's shoulder; your saddle was a flap of moose or buffalo skin or more often the pony's bare back. You rode with the band until the chief gave the sign for a halt. Camp meant a place where there *Indian life.* was a prairie pool or water standing in a buffalo wallow; food meant anything that could be caught, shot, or gathered on the way; sleep meant any spot of ground in a bunch of grass or under a bush where you chose to curl up. To look out for yourself, to ask no questions, and to make no murmur were imperative. In the wild neither man nor beast pays any heed to the cry of the weak. The human, like the wolf, endures in silence.

But the hardships, the dangers, even the *The lure of the wild.* disasters of such an expedition were willingly borne by youth for the love of adventure and the pure joy of the open. That riding day by day, riding into the sunset west, riding into the unknown, had an attraction about it difficult to explain yet none the less real. The unexpected was always expected. No one knew what surprise would be sprung from the top of the next divide. There was always the outlook of the hunter watching for game, for food was the one great demand. And there was also the keen watch of the hunted—the trained eye looking for danger. Advance riders were continually looking over slopes for hostile bands of southern Cheyennes or prowling marauders from over the northern border. The last thing seen at night was scouts riding along the high divides, taking a final look

around the huge circle. What a picture, those half-naked Sioux silhouetted against the blood-red twilight, each one bunched over his pony's shoulder and peering catlike into the gathering gloom! And what nights of mystery with the hushed camp asleep in the great silence —asleep and yet half-consciously listening for the slightest danger sound.

Dawn on the prairie was almost as mysterious as dusk. The light was not shot up suddenly from the bow of the horizon, but gradually grew around the whole wide half-circle of the east and then flooded softly into the upper sky. There was no sharp division-line between the night and the day as in the tropics. The light spread upward through the

dark turning it slowly into half-light. Just as gradually the dark outlines of the divides melted away, some odd spots of shadow around the camp grew into picketed ponies, some moving blurs here and there turned into the Sioux getting ready for the day's march.

An Indian band gets under way quickly, not being bothered with much impedimenta in the shape of baggage. The chief and the hunters rode first, strung out in single file, as elks move when migrating. After them came the squaws, with ponies trailing teepee poles across which were lashed carryalls for the children and the camp plunder. In the rear trotted the dogs—snarling, wolf-like curs that always seem inseparable from an Indian outfit wherever it may be. Only the dogs gave tongue. The band rode on in silence. *The march.*

And silent, too, in the early morning, the wild life of the prairie. Even the whimpering coyote that had yelped at intervals through the night made no whine after daylight. The wolves slipped away to holes under cut-banks or in washouts, the small buffalo ring (cows and calves on the inside and bulls on the outside) that had fought off the wolves through the night broke guard and grazed at ease, the antelope nibbling grass on the far-away ridges were seen only as spots of white flashing in the rising sun. Not a sound from any of them. *Prairie silence.*

Animal life on the prairie.

The prairie was profoundly still save, perhaps, for the call of some lone goose seeking its kind in the prairie pools or the cry of a curlew that fluttered and dipped above the wolf that had come too near her nest.

By ten o'clock floods of light poured from the sky. The tall grass of the lower swales, as it bent and rolled in the morning breeze, seemed to glitter with light, the prairie pools flashed it like mirrors, the gentle roll of the divides received it but projected no shadow.

The heat.

And the heat! Great stretches of it could be seen radiating upward from the bare spots and dry lake beds, the air grew rosy with it, the sky became opalescent. The matted hair of the

Indian endurance.

Sioux and the tough hide of the ponies beat it off; the women, children, and dogs endured it with grim unconcern. The band pushed on. By midday it had, perhaps, picked up a stray buffalo or antelope, the women and children had caught young prairie-chickens or gathered wild onions, carrots, and lily bulbs, and there was a pause in some creek bottom, under a lone cottonwood, for a feast and a doze.

Again the night camp, perhaps at the edge of a long-linked lake, fringed with wild rice, where clouds of ducks came pouring in at dusk, where blue herons waded along the muddy edges, and white swans glided down vast inclines to rest upon the open water. No fear of bog or quicksand could keep the ponies out of it. Into it they pushed, plunging their noses deep under the water, drinking with long gulps, and then lurching back to dry land. And once more at dusk the shadowy tee-pees and ponies, the scouts seen in silhouette against the twilight sky, and the blue overhead studded with golden stars.

Prairie lakes.

The night camp.

Gradually, after many days, the prairie-grass grew shorter and finally disappeared. The buffalo bunch-grass, lying low and curling much, took its place. Gradually the divides grew higher and the swales deeper. Small hills and valleys came into existence. The creeks turned into little rivers with muddy water and steep, sloughing banks, cottonwood grew along them, with here and there clumps of box-elder in which deer and white grouse found a refuge from the noonday heat. The uplands were covered with sage-brush dotted at intervals by clumps of nopal and yucca, the soil had become a "gumbo," the margins of the pools were fringed with a white alkaline sediment

Divides and swales.

The uplands.

instead of wild rice. Moreover, there soon came into being abrupt sinks in the ground—isolated trenches sometimes miles in length with steep banks, sometimes many feet in depth. And to balance them there sprang up from flat plateaus tall, mound-like buttes with domes and turrets and towers rising in fantastic forms. The edge of the Bad Lands had been reached.

Buttes and trenches.

The strange surprises of the Mauvaises Terres—those vast, treeless uplands composed of loose sands, gravels, and sediments, and by reason of insufficient grass or bush covering washed and gulched by every passing rain! The elements of fire, water, and air seemed to have conspired against them. Water was the chief element to drag them down, but the wind honeycombed and rounded them, and fire burnt them. Fires in the underground lignite had been burning for centuries and were still apparent in blue smoke stealing out of faults and fissures in the ground. The burnt buttes showed their burning. Sometimes they were banded and streaked by a hundred different tones of terra-cotta. Strange color was everywhere in that strange land. It was in the sky, the air, the soil. The alkali

The Bad Lands.

Fire and water wear.

Strange colors.

dust stirred by the pony's hoof, when struck by the sunlight, turned into a pale heliotrope, and the pony's moving shadow was lilac-hued.

And what thin, almost breathless atmosphere on those high mesas that succeeded to the Bad Lands! We were on the sage-brush and grass-grown uplands of the Powder River, and the air was so pure and clear that objects at a distance cut out in sharp detail and seemed to move up close to us. Hills twenty miles away were apparently only a half-hour's ride from us, and ten-mile valleys seemed little larger than the prairie swales we had left behind us. At night the air seemed still more wonderful because apparently non-existent. It did not move, you scarcely felt it in breathing, there was not enough of it to make the stars twinkle. The blue Vega and the white Arcturus burned but did not blaze. The Milky Way, with its constellations cast upon a blue-black sky, shone but did not glitter. Oh! the splendor of those soft, windless nights on the Montana table-lands! The smouldering camp-fire, the shadowy teepees, the blurred ponies grazing at a distance seemed the necessary foils to set off that great peace and glory of the sky.

Powder River ranges.

There came an evening when the band drew into a range of hills—beautiful, grass-grown hills with rounded tops that flowed against the sky—and at dusk made camp in a valley at the head of a tributary of the Powder. Before sunrise the hunters were on the rounded tops looking over into a near valley perhaps a *An upland valley.* dozen miles in length. It was not an unusual valley, being treeless and covered with low bunch-grass, like many others. And to the untrained eye there was nothing unwonted in its appearance that morning save that in the far end of it the ground seemed covered for many acres with a brown-black growth. The more one looked at this growth the more *Sighting buffalo.* strangely interesting it became. It was apparently motionless, but along the edges of it there were tiny brown spots in bunches that moved out from the mass of it, and in the central part were other groups or bunches that moved and made gaps or openings in the mass. The brown-black growth was alive, though silent and apparently motionless. It was grazing, milling, pawing, fighting. It was a buffalo herd.

In an hour or so a portion of the Sioux had ridden completely around the back hills and had come in upon the buffalo herd from the rear. The plan of the hunt was to drive the buffalo up the valley in the direction of the camp, where another line of concealed hunters would dash down upon them. Directly the first party came into sight a stampede took place. The huge, brown-black spot got under way with incredible swiftness, forming as it ran a protective line, with the cows and calves on the inside and the bulls on the outer edge. A blue haze of dust arose from it through which one could occasionally see on the outside a Sioux hunter lashing his pony along and shooting as he rode. The small puffs of smoke could be seen but no sound was heard. And occasionally, too, the pursued turned into a pursuer. An angry bull would drive out from the herd with the directness of an arrow and the force of a catapult. Every horse and rider knew the wisdom of keeping away from that monstrous head. The young men, some of them still using the bow and arrow, took risks among the cows and calves, but the elders held a distance and closely watched the bull line.

On came the herd, bumping, rumbling, pushing to get away from the worrying fire of the

The buffalo hunt.

The Sioux attack.

The buffalo defense.

The stampede.

The pursuit and death.

Across the Tongue.

hunters. Then the waiting Sioux, stationed in the foot-hills near the camp, charged down upon the leaders. The great patch, attacked at the head, swerved like a deflected wave, pushed off to the left through a defile in the hills, choked up the defile and rose on the sides of it, pushed over and on into the big valley of the Powder and, with the hunters, disappeared from view. The blue dust gradually settled in the valley, and the only things now seen were brown-black units of the herd, some plunging here and there with broken legs, some tumbling over in attempts to get up, and some lying quite still. Creeping down toward the still ones were gray spots that first reconnoitred, then came close up and set to work. They were the skinners from the camp.

For days the herd was followed and then abandoned for other hopes. The hunter, like the gold-digger, is never content with the present. Just beyond, over in some other river basin, the catch is sure to be easier and more plentiful. The band moved on across the Powder, swam the racing Tongue, and began the ascent to higher table-lands. The high points of the hills soon became fringed with ragged juniper, cedar, and scrub pine; the sage brush still flourished but other upland growths came in; the clay butte did not disappear but rocky ledges began to crop out; white-tailed deer jumped from the bushes and bands of elk scattered with a bump of hoofs through the scrub pine. Around the edge of almost every water hole the paw of the mountain lion could be seen in the alkali mud, and near it the huge clawed foot of the silvertip.

The table-lands.

Deer and elk.

It was a land of game—that high land beyond the Tongue. The hunters were busy from morning until night and the camp poles hung heavy with strings of dried meat. But more wonderful than the game and the growths was the enormous expanse of country that one saw from every high point. Was there ever such a land for distances, for far sight across the hills and buttes and mesas! It was a pink and gold and lilac land, another terra-cotta world lying under a rosy sky. And untenanted. Even the Indian tribes that usually hunted there were not seen by us. It seemed as though

Unten-anted ex-panses.

> "We were the first that ever burst
> Into that silent sea."

Love of the wilderness.

What an exultant throb in the thought! The call of the wild is as strong in the human as in the wolf. He is forever longing to escape from the gilded cage of civilization, longing to get back to savagery. The Gardens of the Hesperides were never so attractive as the uplands of the wilderness.

One day at sunset, at the top of a high divide, the chief drew rein and sat still, looking away, across rivers and hills and valleys, *Far views from the divides.* far away to the south. We came up beside him and followed his gaze. Lying low down along the horizon, perhaps a hundred miles in an air-line, was something that looked like heaped cloud-vapors. It was no more than an outlined mist against the lower sky, a little sharper, perhaps, in the edges than elsewhere; and what made it noticeable to our eyes was its color. In the sunset it had a rosy *Distant mountains.* hue and its edgings were fire-red. Sometimes over arid regions the cloud banks at sunset take on just such color, but these were not clouds. The bronze profile of the chief never relaxed, never showed either pleasure or pain, never showed even so much as recognition of the sight before him. He said something in a low tone but his lips scarcely moved. Was it possible? Was that faint outline of color the high ridge of the Big Horn Mountains down in Wyoming? Could the glow of sunset upon barren rock produce such wonderful color as that?

The Big Horn mountain tops.

The band moved on, rising higher and higher into the short range known as the Cheetah Mountains,* winding on toward the west—hunters, ponies, squaws, papooses, and snarling dogs as before. The marches were now short, the hunting good, the camping easy; but gradually, as the days came and went, valleys were crossed, rivers were forded, new ranges were traversed, until there came an afternoon when again the chief and the foremost riders drew rein upon a lofty height to gaze a moment at a grand panorama outspread before them. It was the look across the basin of the Big Horn River and beyond. Was there ever such another strange vista! A loosely cemented table-land had been gulched by rains and slashed by rivers until thousands of enormous, cone-shaped peaks seemed

The Cheetahs.

Valley of the Big Horn.

---

* I cannot remember what, if any, name the Sioux gave these mountains. On a later trip across Montana, in the early eighties, I found the cowboys calling them the Cheetahs. The name is now geographical, I believe.

pitched tent-like upon the plain. As far as the eye could see this barren and soundless camping ground of nature lay bleaching under a summer sun, the very acme of desolation. The chief gave no sign nor word about it, but the next day the course lay more to the north-west, and, after several campings, we swam the Yellowstone and drew away into a more open country.

More grass-lands, more buffalo, and more hunting. And then a broken, elevated country leading up and out to the west stretched before us. Oh! the beauty of those uplands covered with buffalo and bunch-grass and spattered with pale geraniums, wild roses, lavender daisies, and sago lilies! As we rose to the tops of the ridges, what beauty in the long, swinging lines of the divides! These

were the high and wide benches that led up to the Rockies—huge green-and-golden mesas spread out in apparently endless reaches. What air and sunlight, what glorious color, what a feeling of elation and aloofness on those camas-blossomed table-lands! In its primitive, untrodden state the sun never shone

on a more beautiful world. The reach of it in tremendous steps leading down to the Missouri was so great that no one could imagine it other than a whole world in itself. And never a thought of any other existence occurred to one. All space was ours, time was not, and history, civilization, and the arts were flung behind us like broken baubles of the past.

We were rising step by step, mesa by mesa, to the great divide, the sky-line of the continent, the high range of the Rockies. At last one day some peaks of the range swung into view. They were far away, a blue-and-silver wonder, and two or three of them were as chalky white as the cliffs on the Powder River. It was difficult at noonday to persuade one's self that they were not mountains of chalk. But their bases were too dark, their tops too light for deception. The white cones of the central peaks were unmistakably snow caps. What radiant beauty in those distant peaks against the blue! The sunset fire on the Big Horn Mountains had not the serenity nor the purity of the snowy crowns. For days they hung upon our horizon. We looked to them in the morning and saw them glittering white, at noon we saw them grow pale and shadowy under the midday heat, and at night the last shaft of sunset fire was reflected rose-hued from their summits. Seen through that thin,

The distant Rockies.

Snow peaks.

Sunset on the peaks.

untainted air, in that wild, untrodden land, the high peaks of the Rockies had a peculiar glory of their own.

The return. The Sioux band turned sharply to the north, swung in an enormous circle to the east, and began hunting homeward. They were seeking buffaloes, not mountains. The white peaks sank below the western verge and were not seen again for many years, the table-lands and foot-hills fell away into the Bad Lands, and the prairie came back to us. The hunt was over. We were once more on the savage edge of civilization.

Memories. But what memories of the wild remained! Brightest of all was the vision of the far-away white peaks of the Rockies. Many ranges in many portions of the earth—Mexican sierras, Spanish cordilleras, Alpine peaks, and even "the frosty Caucasus"—have been seen and studied since then; but that sight, seen in the early days from the uplands of Montana, has *First im-* never faded, never been equalled. It was a *pressions.* first impression of the high mountains and has remained through many years the most lasting impression of all.

It is something of a pity that the Dacotah prairies and the Montana uplands have undergone such great changes. The sea of grass that stretched away far westward to the setting sun, the prairie flowers, the buffalo, wolves, and Indians have all been ploughed under, and ridges of grain are now growing where the antelope grazed and the wild horse stamped. The wilderness is no more and the blossomed plain but a border memory. That civilization replaced the prairie with the farm may argue something for human well-being, but the change has been at the expense of natural beauty. Nature always goes out when the white man comes in.

The great change in Dacotah.

The vanished life.

Fortunately the higher peaks of the Rockies are still practically inaccessible. The white crests lift aloft as though defiant of change. Their walls of rock and peaks of snow may be seen a long way off, looking up from such comparatively wild country still as the Montana table-lands. There they seem to stand as a barrier, a parapet against any further invasion.

The mountain wall still stands.

> "Long may they stand—the ages through—
>   Undimmed by time, untrod by man,
>   A flash of light on Nature's plan,
> A wall of white against the blue."

CHAPTER II

MOUNTAIN-MAKING

Indian dislike of mountains.

The love of the open.

THE Indians were never too fond of the mountains—in fact, they cared less for them than for the prairies. Occasionally they went into the Big Horn to hunt mountain sheep or they pushed after elk up into the high foothills of the Rockies; but they never cared to live there. They disliked the timber and preferred the edge of open country such as the Bad Lands. In this they were not different from the furred and feathered nature about them. There are few birds in the mountain forests as compared with those in the open; and what animals live there do so by force of circumstance. They go there seeking protection. No doubt the sheep and the goats now love the semi-barren regions of the mountain above the timber-line, having adapted themselves to them; but they went there originally under compulsion, and they live there now largely as a condition of living at all.

With the Sioux there was always some superstitious fear regarding the mountains. They believed that the Great Spirit rested upon the summit of the Rockies and that sometimes in the evening glow he sat there with the white feathers of his war-bonnet streaming in the west wind, smoking his red pipe-stone and sorrowing over the disappearance of his people. All the lesser peaks were peopled by evil spirits who were seeking to destroy the Great Father. Wakon-Shecha, the Evil One, had his abode in weird canyons and deep mountain caves, and came forth by night seeking whom he might devour. The younger men laughed at the tales, but they usually took good care to come away from the mountains before dark. They felt safer in the open country.

Their fears were groundless, to be sure, but by no means singular. Other races, with larger frontal development than the Sioux, have believed similar things about the mountains. Indeed, it seems to have been an early superstition of mankind in general that the mountains were peopled with gods, that they were even created by them. In the far-back heroic days the mountains had been hurled hither and yon in Titanic combat; and the little hills were dropped as landmarks from the

Siva the creator.

Sacred mountains.

lap of a huge, striding giantess that she might know her way back home. In India Siva was the creator and shaper of the world. He dwelt in the remotest heights of the Himalayas enwrapped in purple-and-gold clouds and seated on a bed of snow-white glittering diamonds. It was from mountain tops that Buddha and Christ ascended into heaven; from them Moses received the tables of the law and saw the promised land; and in the heights of the Caucasus Prometheus was chained. Almost every one of the older countries had its sacred mountain, from which the gods came and went —Lofen in China, Fujiyama in Japan, Samanala in Ceylon, Ararat in Armenia, Lebanon in Palestine. In Greece each separate mountain seemed to have its presiding deity. Pan was in every mountain grove; Apollo and the Muses held revelry on the top of Parnassus; while Zeus and his following heard the hymns and prayers of the faithful from the heights of Olympus.

At the north the tale varied. A Scandinavian belief prevailed that the mountain contained a hoard of gold guarded by gnomes and frightful dwarfs and that the queen of the mountain, sitting in state, was a virgin of ice who turned the bridegroom beside her into everlasting crystal. In Germany the queen of the mountain was Venus and those who entered her court were doomed eternally. Tannhäuser, by the aid of the Blessed Virgin, was the only one that ever escaped from the Venusberg. At the entrance could be heard the groans and lamentations of those who had ventured within. In the Harz, on the summit of the Brocken, all the witches of earth and air assembled on Walpurgis Night to enjoy high carnival. The mountain heights were peopled thick with myrmidons, pygmies, emmets, fingerlings, furies, spectres—all the unearthly rabble that usually travels in the train of Mephistopheles. Weird lightnings flashed along the Blocksberg and thunders shook the ground.

Teutonic beliefs.

The Brocken.

In later times, though there was a persistent faith in the evil spirits of the mountains, there also came in again a belief in good spirits who dwelt there and ruled the destinies of men. There was a renewal of

Good spirits in mountains.

"The fair humanities of old religion,
The power, the beauty, and the majesty
That had their haunts in dale or piny mountain."

Shrines to the Deity were placed in the high hills as though they were the world's great altars. People went up in the mountains to pray, seeking a sign from the heights in a pillar of fire or a cloud of storm. St. Francis and St. Catherine, with the long list of Christian hermits, anchorites, mystics, dwelt in the mountains, and in their names convents, chapels, and monasteries were builded along eerie ledges and beetling promontories—fair fanes of prayer that seemed to be nearer God in that high, thin air.

For before ever the mountains were brought forth, or ever the earth and the world were formed, from everlasting to everlasting Jehovah had been. He made the mountains and fashioned the earth and covered it as with a garment of blue sea and sky.

But the belief in mountains as gods, or the abodes of gods, has quite passed away in modern times.

Science has changed all that and furnished us with a more prosaic creed—one that opposes the old. It was to be expected that human thought, after arguing for centuries that the world was made in six days by a personal God, should finally fly to the other extreme and insist that the world was made in six hundred million years and that it made itself. The antagonism of science to religion was violent; the departure in the opposite direction was radical. And yet, notwithstanding the cautious might be inclined to reject both beliefs, it is impossible not to pin some faith to the conclusions of science. At any rate there is no understanding the modern theories of world-building and mountain-forming except by paying heed to the finger-prints of geology.

World-building.

It seems that science eventually grows weary of its own hypotheses and discards them from time to time quite as readily as it does the conclusions of religion. The nebular hypothesis of La Place was true so long as no one gave it serious thought, or found it in conflict with newer theories; but now it has been modified and practically superseded by a planetessimal theory put forth by Professor Chamberlin. The new hypothesis certainly seems more plausible than the old. It assumes the origin of our planetary system in a hot, gaseous body which, approached by some large or controlling star, threw off from itself protuberances beyond its own controlling gravity. These protuberances of gaseous matter became planetessimals, or scattered particles of a nebula—the common

The nebular hypothesis.

The planetessimal theory.

white nebula of the telescope, spiral in form, with an outward rotary motion from left to right, and with a large central mass or sun. The nebulous matter of the spiral was in a finely divided state (as the continuous spectrum implies) and came to that condition probably by a process of cooling. The particles of it, both great and small, revolved in elliptical orbits about the central sun and gradually formed in aggregations or knots along the arms of the spiral, as shown to us by the camera. These knots were the nuclei of planets and it was to them, or to the sun itself, that the remaining diffused and scattered particles of the

nebula were eventually added. The mode of addition or accretion was not by simple gravitation but by collision and conjunction in orbital motions. The planetessimals or flying particles had each an elliptical orbit of varying eccentricity about a common centre, and they were all finally gathered in by "the crossing of the elliptical orbits in the course of their inevitable shiftings."

Out of this process the planetessimal hypothesis develops a solar system such as ours, gives consistent views of the requisite distribution of mass and momentum, of the spacing out of the planets, of their directions of rotation, of their variations of mass, of their varying densities, and of other peculiarities. In a sentence it accounts for the formation of the earth by a system of planetessimal accretion instead of postulating a molten liquid world, revolving in space, and gradually cooling from the outside toward the centre.

The theory, therefore, seems to combat the old idea that the earth has an interior of molten fire and that we are living merely on its thin outer crust. It postulates the gradual accretion of planetessimals, the upbuilding of the world by slow additions, and the solidifying of the globe at the centre extending and cooling outwardly as it grew. The internal heat of it—sufficient to mould it into spherical shape—was largely the result of after-compression through gravity. During the growth of the earth, as to-day, large amounts of heat were carried off by volcanoes. The source of this heat is thought not to lie in the centre of the earth, but in a middle zone where the compression would be mathematically greater. And in that zone also are supposed to be the sources of earthquakes.

This theory of the origin of our solar system

Stages of growth.

is not vital to an understanding of the earth's subsequent history. We need not start so far back, and still perhaps gather some notion of how the mountains were formed. If we begin with a cooled but volcanic globe, the after stages of growth, whereby it acquired irregularities of surface, may be traced with some certainty. There seem to have been several of these stages, covering enormous periods of time. The first was one controlled

The accession period.

by the accessions of planetessimals to the exterior. It is possible that some of the protuberances of the earth, some of its surface deformities, are due to the irregular manner in which the infall of planetessimals took place. It is possible, again, that the globe never at any time became a perfect sphere and that its present broken surface is its original condition less modification by the

Heat and gravity.

elements. Geological science, however, does not admit such possibilities. It postulates for this first period only increasing heat and gravity.

Vulcanism.

The second stage is that of vulcanism. The heat arising from compression, collision, and molecular rearrangement resulted in some melting in local spots, in arrested ascensions, gas formations, explosions, volcanic eruptions. This transfer of molten material from the interior to the exterior in the great volcanic age was no doubt something of a factor again in forming the irregularities of the earth's surface; but vulcanism in itself is not sufficient to account for sunken sea-beds, lifted continents, and folded mountains.

The third stage — the hydro-atmospheric eon—seems responsible for the continent and sea-bed making and the consequent readjustments by the stretching and wrinkling of the crust. It seems that when the water vapor of the atmosphere by cooling finally reached the saturation-point it took liquid form. This, with some underground condensation, made up what is known as the hydrosphere. The water came out of the atmosphere, and up from the ground, and settled in the depressed volcanic pits. These grew and became crater lakes, and, finally, by aggregation and depression, they became shallow seas. The water-covered areas acquired higher specific gravity than the land areas, the settling of sediments in the basins increased the weight, the sea-beds gradually subsided. As the sea-beds sank the continents arose.

The sinkings of the ocean-beds were the master movements, not only in what they left exposed in high peaks and ranges, but in what they forced up or bent or twisted as they sank. No doubt these movements were the cause of many uplifted areas, of the stretching of the continents, of the thrusts and many small foldings and wavings of the crust; but it seems that they do not account for the formation of ranges like the Alps or the Rockies. They were doubtless contributing causes, for not one but many movements have combined to make the mountains what they are to-day.

Among them the wrinkling of the crust through age and the cooling of the earth must be reckoned with.

It is an accepted theory that the earth as it cooled contracted in volume and that the outer shell of it warped and buckled into waves and folds. Apparently these foldings

wrinkled over the continents in mountain ridges and under the seas in abysmal depressions; but in reality perhaps the continents and sea-beds were caused by the contraction of the globe as a whole, or at least of the whole crust of it, whereas the mountains may have been formed by foldings of the outside layers

of the crust only. The first would be a matter
of huge lateral thrust and vast readjustments
of the entire earth surface; the second would
be a matter of folding ranges, snapping arches,
and splintered peaks over limited districts.

Such at least, in brief, is the hypothesis regarding the making of the high mountains
put forth by Chamberlin, Suess, and others,
and largely accepted by the modern scientific
world. It does not account for everything,
and, besides, it requires belief in some rather
arbitrary figurings. For instance, the crust
of the earth that by wrinkling produced the
Alps and other high ranges is mathematically
computed to be only an outer layer of, say,
three to five miles in thickness. That sheer
zone seems unbelievably thin. And, again, the
explanation of the rise of old sea-beds into
mountain heights is not very satisfactory unless it be postulated that the earth was once
a perfect sphere, and entirely covered with
water. The modern scientists are inclined to
think that the general movement has been a
sinking one, and that the upward movements
or foldings have never equalled it. And yet
it is very certain that many of the high mountain ranges are formed of sedimentary rocks

Faulting.

*Exposed
face walls.*

*Elevations
and de-
pressions.*

that were once under the sea. The theory is not entirely free from objection.

There is a second method of mountain-making of which geological science takes cognizance, but its workings though common are not apparent everywhere. This is mountain-making by "faulting." Certain cross-sectional breaks in the earth's crust allow portions of the crust to sink and in doing so push up, or dislocate, or, at least, leave exposed in sheer walls other huge portions. It is as though a stone arch had been disrupted and some stones of it forced up while others were permitted to drop down. Occasionally this dislocation, with an exposure of face walls, takes place on a grand scale, as in the Vosges or the Oregon mountains, and sometimes its presence is not suspected save by geologists. The plains of Lombardy, for instance, lying under the Alps, are supposed to be the result of faulting. Often a great elevation is accompanied by a consequent depression of a near surface. It is even conjectured that the white Alps and the gray Carpathians have had something to do with the depressed Mediterranean and Adriatic south of them. But about that there is no certainty.

A third method of mountain-making is recognizable in vulcanism. Volcanic mountains are simpler in construction than those in the ranges. They are merely the results of accumulation. The volcanic cones that we have seen grow from decade to decade are uniform in shape and have no wrinklings or foldings of rock about them. Colima, Kilauea, Vesuvius, Etna, and their kind, are the blow-holes of an overheated under-surface, and their ashes and cinders, while formidable aggregations close to view, are comparatively unimportant on the earth's surface. That volcanic action and lava beds were once more prevalent than now is highly probable. There was a time when possibly the whole surface of the earth was pitted by volcanic craters. But vulcanism is not thought sufficient to account for the surface deformations of the globe that we know to-day.

Volcanic mountain-making.

Cones and craters.

The theory of a century ago that all mountains were produced by upheaval no longer obtains. Science has reversed itself and decided that subsidence and compression rather than upheaval caused the various ranges. And yet upheaval would explain range formation quite as readily as the mole explains

Upheaval as an explanation.

the ridge across the lawn if the heaving force (a large enough mole) could be discovered. Is it not possible that vulcanism may be that force? Why is it necessary to assume that the heat of the interior of the earth manifested itself in only one way—in the spectacular spouting volcano? Suppose there were at one time rock-fissures filled with half-molten lava lying under what are now the Rockies and the Andes. Suppose that through sand or gravel beds, or along broken rock-cleavages the waters of the Pacific seeped through to these underground beds of lava.* Would the steam generated be sufficient to lift the mole-ridge of the Rockies without, perhaps, breaking through the overhead arch and escaping into the atmosphere? Might not the steam follow the line of least resistance along a disrupted rock ledge perhaps for hundreds of miles, straining, heaving, crackling the beds above it without escaping entirely?

For a wrinkle of the earth made by lateral

* The volcano Cosequina, at a late eruption, covered the ground about for twenty-five miles with muddy water sixteen feet deep in places. Stromboli throws out vapor every fifteen minutes, and many other volcanoes show the presence of steam in enormous quantities.

thrust the Rockies seem too large, and too arbitrary in form, while the Alps and the Caucasus are too concentrated. The Alps, for instance, if flattened out would make at least seventy-five miles of crust. Perhaps it was a lake or sea of lava under the Alps instead of a long-fissured bed that was set in action by Mediterranean water, as possibly the Caucasus by the Black Sea and the Caspian, and the Himalayas by the Indian Ocean. And how shall lateral thrust account for such round and isolated eruptions on the earth's surface as the Black Hills? Are earthquakes —and they happen almost always along the seashore—caused by muffled underground explosions of steam and lava? Was the internal heat greater in the mountain-making era? And what is now merely an earthquake—was it then a direct upheaval into mountain ranges? Were the causes the same but the later effect reduced by exhaustion of power? Science gives negative answers to such questions.

Making of the Alps.

Lateral thrust insufficient.

A fourth method of mountain-making, spreading over a wider field of activity and with far greater results in tearing down if not uplifting the earth's surface, may be found in erosion. Mountains have been carved out

Effects of erosion.

Water wear on buttes and hills.

from the earth's crust, valleys have been widened and sunk, and huge canyons like that of the Colorado have been cut from the solid rock by water. Wherever the watercourses run beds are sunken and shores of high or low degree are left standing, sometimes in the form of bluffs, as along the upper Mississippi, and sometimes in rounded mountains, as in the Catskills. The buttes of the Bad Lands, with their fantastic forms, are merely the result of erosion; and many of the high mountains of Colorado are hard cores of stone from which water has washed away the softer surroundings. Originally, perhaps, all these regions were elevated plateaus heaved up by the sea, and were then (starting with an uneven surface) beset by winds, rains, and running streams until ranges of hills were left prominent by the washing out of their intermediate valleys.

Mountain disintegration.

Even the high mountains, heaved upward by tremendous lateral thrust—mountains like the Himalayas—are subject to great modifications through erosion. They are no sooner made than the elements—water, wind, snow, frost—begin tearing them down, denuding them, disintegrating them. The earliest stream-beds running down the mountain sides were possibly the beginnings of what are now known *Early stream-beds.* as the transverse valleys. The streams carved out the valleys, cut through the passes, rounded the foot-hills, and by their sedimental deposits made the plains. Just so the intermediate valleys of the parallel ranges were the courses of the larger streams and the beds of the long mountain lakes. All waters are agents of denudation and are engaged in dragging down the high ranges to the level of the flat plains.

That mountains of tremendous proportions have been, and have passed away from the face of the earth, there can be little doubt. Under the grass-grown plains, the level lands, *Worn-down mountains.* one finds folds in the rock strata indicating plainly enough the roots of mountains now worn to the level and hidden under vegetation and river deposits. The half-way stage of destruction is apparent in many of the low hills and table-lands—in the round mountains of Saxony, the Highlands of Scotland, the uplands of Bohemia. The high mountains where destruction is now going on so fiercely are the newcomers on the globe. The sharp edges of *The Alps and Andes new mountains.* the Alps and the Andes suggest their comparatively recent origin. Eventually those mighty ranges will be rounded like the Tuscan hills or flattened like the Arizona mesas. Nothing endures. The "everlasting hills" is awe-inspiring and poetic, but it is not true. Change in every aspect and feature is the law of the globe itself, as of the life that lives upon it.

CHAPTER III

THE HILLS

IF we accept the explanation of the earth's structure offered by geologists we shall have to abandon some of the theories advanced by the more enthusiastic nature writers. Mr. Ruskin, for instance, is very right in pointing out that the mountains with their ragged, upright lines give a feeling of action, and that the plains by their flat, horizontal lines give the opposite feeling of repose; but he is misleading in likening the mountains to the violent muscular action of man "brought out with fierce and convulsive energy, full of expression, passion, and strength." It is a false analogy to say that "the mountains are the bones of the earth." And to intimate that they have risen by initial energy into "vast pyramids or wedges, flinging their garment of earth away from them on each side" is to put forth an untenable conception of mountain-making.

Ruskin on mountains.

False analogies.

Whatever feeling there may be about action in the mountains and repose in the plains is purely subjective with us. The ranges themselves are passive masses of inorganic matter pushed into place, or accumulated, or carved out by compelling circumstances. As for muscle and bone structure in or about the mountains they never had existence. The long sierra running up and down the western continent is not the spinal vertebræ of the globe nor is it a place where any extra strength is put out either for defense or display. On the contrary it is a distinct line of weakness.

The globe is supposed to be covered by a thin outer shell, somewhat like that of a hazelnut, perhaps, and the mountains are simply cracks or lifts or faults in the shell. Strength in relation to the rest of the shell is not their quality, and yet because of their enormous height and mass they do produce the feeling of power with us. A slash in the earth, like the Grand Canyon of the Colorado, creates a similar impression. But we should not be misled by either of them. They are uplifts or washouts, points of fracture or wear in the shell, and not layers of extra granite placed for brace or support.

Quite as misleading as this idea of mountain strength is the fancy that the low hills are the buttresses or supports of the higher ridges. *Low hills as buttresses.* Mr. Ruskin has it that "the masses of the lower hills are laid over and against their sides like the masses of lateral masonry against the skeleton arch of an unfinished bridge, except that they slope up to and lean against the central ridge." Once more there is a superficial truth in this—a truth of appearance. *Truth of appearance.* Erosion has worn away the lower hills at the top and extended them outward at the bottom so that, where there are no intervening valleys, they have the look of leaning inward toward the central range. But the appearance will not bear analysis. Constructively, the foot-hills do not "lean against" the high ridges any more than the little wrinkles in the skin of a withered apple lean against the big wrinkles. The foot-hills are usually the points *Foot-hills.* where the lateral pressure has been less violent and as a result the buckling is less marked— that is all. As for "support," a pyramid is usually upheld by its base, but a mountain is not always in pyramidal form. Sometimes it comes down by a sheer wall or precipice to the plain or the sea without intervening or sloping hills, as witness the front range of the Rockies in Colorado. And, again, the lower

Isolated hills.

hills are often some distance removed from the central ridges. They may be a group quite by themselves supporting nothing, as, for example, the Black Hills, or they may be in detached localities as with some of the hills in the American and Mexican deserts.

Buckling of crust not uniform.

The truth is that the plains, the foot-hills, and the high mountains are all parts of the one crust and they have undergone various stages of rest or upheaval just as the lateral pressure of the contracting shell has lifted them or neglected them. The precise form, or even the general form, of the buckling cannot be laid down arbitrarily. That is to say, there is no rule of mountain-making—no one form to which they all pay allegiance. It is an error to suppose that one range is formed like another, and that when you have seen one you can generalize about all. The

The Alps peculiar.

Alps, from which most of the generalizations have gone out, are perhaps the most abrupt and violent of the mountain groups and are more of an exception than a rule. The Mexican mountains, with their central ridge running for hundreds of miles, are quite different. Again, the foot-hills of the Vosges are no criterion whereby one can judge the foot-hills of the

Rockies. As for the plains, some of them lie low down and flat like those of Lombardy or Hungary, while others are high up and broken like those of Tibet or New Mexico. *The undisturbed plains.*

Yet different as they may be in height and formation, there still remain these three phases of the continental crust—the mountain, the hill, and the plain. Each class within itself must, of course, be regarded in its various manifestations, for, again, there is no general description that will fit a plain or a hill more than it will epitomize a mountain. Among the hills, for instance, every county in England will produce something a little different from every other county, and each state in America from Massachusetts to California can boast the unique form or color or bulk of its particular group of hills. Nevertheless, there is a tendency on the part of nature, after she has given birth to the hills, to fashion them by erosion into one common form. That form is somewhat like an inverted bowl—an arch lifting toward the zenith. This upward curve of the hills is offset by valleys that sink bowl-shaped in the opposite direction toward the centre of the earth. The hill alternates with the valley, and the flow upward and down- *Variety in the hills.* *English and American hills.* *Forms of hills.*

ward is apparently as rhythmical as sea waves in a Greek decorative pattern.

But whatever the tendency in hill and valley building, there is no such uniform appearance as we have suggested. Many factors come in to modify or change the smoothness of the domed hills, such as faults in the strata and water wear, which produce gorges and canyons, cliffs and bluffs, ravines and hollows. The hills that closely approach the bowl shape are composed of gravel or loose silt. They are frequently cut out of old sea-beds, as in the case of the Montana buttes, or have been deposited as the residue of glaciers. They have no rock strata about them and hence are easily washed by rains into the form of a knoll or knob or mound. Usually they are guttered on the sides and in the course of time flatten down to the level of the plain. They are infrequently met with and are not precisely what we mean by "hills."

Nor is the volcanic ash, heaped in a cone and afterward denuded by erosion, any nearer approach to a hill. This form of mound is seen in many parts of the world, is generally without vegetation, and is usually gray and forbidding in appearance. The lava and sco-

Gravel hills.

Water wear upon them.

Volcanic hills.

riæ of it make a hard kernel difficult to wear down, and perhaps it will stand against the elements for centuries because of its fire-hardened contents. Eventually it goes the way of all earth-protuberances, that is, breaks down under water wear, perhaps gathers unto itself a crater lake, and becomes merely a guide-book curiosity. It is never at any time the kind of hill we associate with a livable or lovable landscape.

Crater lakes.

The rock beds that have been heaved up, but without sufficient lateral thrust to cause snapping at the top of the arch, are perhaps of commoner occurrence. Their stratification is not easily recognized unless there are breaks in the surface. When such exposure takes place the rock-layers are usually found to be of sandstone, limestone, pudding-stone—strata of a sedimentary or secondary origin. The granites and gneisses—the harder igneous rocks—do not usually appear in these hills, though there is no sharp division-line in the appearance of soft and hard rocks, here, nor there, nor elsewhere. The quality of the rock, however, often determines the shape of the hill and the shape lends a clue to its identity. The softer sandstones with clays and gravels

Hills of rock for-mation.

Soft and hard rocks.

produce something rounded, as the harder granites something abrupt or precipitous.

Limestone beds.

The limestones lie in beds sometimes many feet in thickness, and occasionally these beds break through from top to bottom like a sun-dried brick. When one part of the brick sinks down and leaves the other part exposed to view, we have what is known, geologically, as a "fault." This faulted structure produces a variety of hill familiar enough in almost every land. It is the hill with a sloping back and a precipitous front. The face wall is often sheer. From it water drips and mosses cling and swallows come and go in flights, and at its top local government or benevolence builds an observatory from which one may look out across a plain at a "view." It is not usually a high hill, though lofty mountains in plenty have been made by this faulted structure. Its vertical face wall produces an exaggerated idea of height and perhaps deceives us. Often the rounded hill is quite as high, but its gradual slope upward minimizes its altitude.

Faulted structure.

The vertical face wall.

Fate of the face wall.

The precipice is not exclusively of faulted origin. The cutting of a stream frequently leaves sheer walls and rocky bluffs exposed, as one may see at the present time in many places up and down the valleys of the Mississippi and the Hudson. In its later stages the abrupt wall disappears by weathering, the bluff becomes rounded, the stream-bed spreads into a valley, the canyon into a clove or a notch, and we have such rounded hills and mountains as those of the Catskills and the Adirondacks.

Rounded hills.

In fact, the rounded hill with which every community is more or less familiar is the every-day illustration of erosion—water wear, wind wear, ice wear. The hills of England and Scotland, now covered with grass and heather, are perhaps the most typical examples. Else-

Effects of erosion.

where, in arid regions where there is no vegetation to stop the rush of winds and rains, strange forms are often produced by this sculpture of the elements. The mushroom-shaped knoll of rock, the enchanted mesa, the organ-piped summits and honeycombed foothills are frequently met with in every desert region.

Worn-down mountains.

Nine times out of ten the rounded hills are but worn-down mountains—old roots of ranges that once lifted skyward with needles bristling like the Dolomites, but are now, after centuries, merely smooth fragments of their former selves. Mountains beaten to the plain by the elements! Such is the common fate of them all, no matter how diamond-edged the pinnacles, nor how thick the strata, nor how massive the guarding walls. The Appalachians and the Tuscan mountains, in their lower spurs, are the mute witnesses of change through century after century. Many other ranges on the face of the globe bear similar testimony. *Processes of renewal.* It is nature's plan throughout all of her works to build up with one hand and tear down with the other. Death is a process of renewal akin to life. The new comes out of the old. Worlds, and upon them mountain ranges, are born and die and are renewed not unlike humanity.

Sky-lines. The worn-down mountains turned into hills that make the flowing sky-line of our horizon have always caught our fancy and held it longer than the snowy peaks. We live with them, know them, love them. There is some physical as well as sentimental reason for our *Effect of lines upon the eye.* liking. The abrupt line of a splintered mountain ridge checks and harries the eye in a measure as a cross-cut saw the hand that passes over it; but the rounded line of a gentle hill— Hogarth's line of beauty—soothes and pleases the eye by its rhythm, its continuity. There is no sharp opposition about the latter. It is true that the hills and valleys do not flow from one to another with the uniformity of sea waves; yet, nevertheless, it is the tendency of erosion in its first stage to model the hills into crests and billows and finally to beat them down to a flat sea surface. In the modelling process lines of wonderful beauty are evolved, lines as serpentine as those of water, and, indeed, formed by water—the most graceful line-maker of all the elements.

The obvious among these hill lines are the edgings or outlines. We see the hills in relief against the sky, and their rounded contours seem to pass from one into the other without a break in the rhythmical sequence. They curve and interweave along the sky and down through the shadow of the valleys until the flattened silhouette becomes a beautiful tracery of greenish purple against golden blue. All low hills are outlined with curves that catch up and carry on one another, that repeat and strengthen one another, that reverse and renew one another. The sweep downward of a hillside is continued in a sweep outward into the valley of the talus; the descending lines on either side of a ravine meet at the bottom

Edgings and out-lines.

Valley lines.

of the ravine and flow downward into the valley again; along the hill bases every deposit of silt or débris from a stream spreads in the lines of a flattened fan upon the valley; and along the hilltops every arch of soil or rock has the lines of an upright fan spread against the sky.

Sparsely covered hills.

Now, these lines of beauty are shown to the best advantage only on sparsely covered hills —that is, hills devoid of thick brush or trees. The heather of the Scotch hills or the grass of the English hills along the sea does not perceptibly check the flow of the lines, but the trees of the Harz or the Catskills muffle and confuse. As soon as timber covers the slopes the lines are softened, weakened, perhaps de-

Foliage muffling hill lines.

stroyed. It is possible in sculpture to place drapery over the human figure and make it reveal the very thing it covers; but you cannot have foliage covering the hills and still show through it the rock structure or the earth curves beneath it. Of course, almost every Englishman, with Mr. Ruskin and Mr. Hamerton for prophets, despises the bare hills and is always talking about green trees, pleasant shadow, and sunlit meadow, as though all landscape should be judged by the city park.

He has, perhaps, never seen the bare hills of
Greece in summer or the barren mountains
of Morocco in spring; he does not care for
lines in landscape, but only for color, light,
and shade.

But the drawing of the hills is their most
commanding feature. And the lines become
more expressive with their severity. The
heather and grass grown hills are wonderfully graceful, strangely beautiful in combination and pattern; but the angle lines of the
splintered peak, the ragged ridge, the notched
descent, though they may worry the eye and
are eloquent of desolation, nevertheless reveal mountain force and character. The
strength of the hills is theirs and with it some
of the sublimity of the higher ridges. Not
only the lines but the surface of bare rock produces the feeling of permanence, endurance,
and power. Every cowboy who rides the open
foot-hills of the Rockies is impressed by that
feeling. He will not reason it out æsthetically
but he knows it instinctively. The bare hills
are his delight and the timbered slopes are his
aversion. It is not that the one is merely
easier to ride over than the other. He feels
in the Arizona hills, for instance, the mass

and majesty of the earth, but in the New Jersey hills only a graceful effect which he perhaps dislikes for its weakness. And, in degree, he is right in his conclusion. The bare hills and ranges are the most powerful features of landscape.

For, be it observed, that the feeling of power and permanence in the hill lies not alone in the outlined pattern against the sky but in what that pattern reveals of mass. There is something more than mere height and breadth. The third dimension (thickness) is also there. The depth through, the body and bulk and weight of the hill, are inevitably suggested in the outline. How impressively and powerfully such painters as Courbet and Winslow Homer have shown this mass of the bare hill and the mountain base! They have caught the outline in its abrupt strength and modelled the surface into positive bulk. Again, Hobbema, Cuyp, Wilson, Homer Martin have used the outlined hill in the foreground of their pictures, as a barrier whereby they could suggest great distance back and beyond it. And, still again, with Rembrandt and Hercules Seghers, the dark wall of the hill, whether cut diagonally or horizontally across the foreground, was used to emphasize the penetrating light and profound depth of the sky above it. The rock barrier, in nature as in art, not only reveals itself, but, by contrast, the beyond of light which it may be placed against and partly hide.

Rock barriers and light.

But, with all the strength of the barren heights, it is impossible to deny the charm and loveliness of the grass-grown or close-covered hills. The heather of the Grampians in early September is the added note of color splendor upon fine form and beautiful linear pattern. And how beautiful, beyond all others, perhaps, those Scotch hills mingle and cross and weave their graceful lines!

The Grampians.

"Fair lines of beauty made by light and shade
    In cadence swinging like an ocean swell,
  Now bright with purple, now with silver grayed,
    How gracefully you charm us with your
      spell!

"You pass, you vanish, in the dusk of night,
    You faint and swoon beneath the noontide
      glare,
  You are mere tracery of broken light,
    But while you last how beautiful! how fair!"

Surely the purple tapestry of the hills was never more brilliantly shown than here! And just so, with perhaps less splendor and more refinement of color, in the sparsely grown grass-hills of Hampshire and Kent bordering on the Channel. Their coloring is a thin staining of the canvas only, but it helps out the drawing. The lines are but the edges of pale fields of color that flow from tone to tone, but they are sufficient.

In America more brilliant effects are seen when the hillsides are spread thick with daisies or wild asters or golden-rod, but the hill lines are not so well shown under the taller growths.

A more pronounced note of color comes to the hills when they are covered with the yellow and scarlet foliage of autumn.

Those who have not seen the brilliant mantle of the Catskills or the New England hills in early October hardly know to what a height nature carries her gamut of hues. Nothing in Europe is comparable to it, and our English friends, who dearly love their blue-green Constable landscape, can only gasp and wonder over our panorama of carmine and gold. But their objection that our autumn hillside is "too loud" finds small acceptance.

Nature knows how to stir the whole of her glowing palette into harmony as readily as to beat accord out of a few sombre notes. The great half-tone of autumnal light and air cements and binds all colors no matter what their hue or quality.

Hills in winter. With the winter comes a great change in the color scheme. The trees are stripped of leaves, the flowers are gone, the grass lies matted and dead, the boulders push out of the ground, and the hillside changes its curved lines for broken and abrupt lines. It is the season for drawing more than color, and yet what delicate grays, silvers, and lilacs are noticeable at just this time! One wonders occasionally if the flare of autumn is not outdone by this precious livery of silver and gray. It is so refined, so serenely beautiful! *Snow upon the hills.* With the coming of the snow there is still another change. The color scheme is weakened by a great admixture of white and the linear pattern is very much cut down by the long curves of fallen snow. These snow-lines, especially when made by the wind, are often quite perfect, as we shall see hereafter; but they are not strong, and the hills are muffled or, at best, shown in swollen or lumpy proportions under the white drapery.

The bare hills of the desert that know neither

Desert hills.

summer foliage nor winter snows are often the most perfect in color of them all. The terra-cottas, the porphyry reds, the pale lavenders, the dull chromes, the rose-madders flush the rock surfaces at every turn, and the peculiar light and gas-blue air of the desert give them all a weird and hectic bloom that baffles the painter's palette to imitate. It is astonishing that bare rocks can reflect so much splendor, that nature out of inorganic matter can wring

Color of desert hills.

so much color that suggests organic life. True enough, it is the color of disintegration and is brought about by decay, just as the rotting of glass produces its iridescence; but the color is beautiful for all that. And what a never-ending wonder to some that nature should lavish so much precious pigment on the barren hills! Doctor Johnson thought that the moun-tains were so much hopeless sterility "dis-missed by nature from her care." The cock-ney limitation of that thought is amazing.

Peaks of the Colo-rado Des-ert.

What would he have said had he seen, all in a summer glow, the red, opalescent peaks of the Colorado Desert or the heliotrope hills of Athens that frame the Parthenon!

Oftentimes the low hills as well as the moun-tains seem little more than reflections of sky splendor, protruding points of the earth designed to catch high light and color. At dawn what a target they are for the arrows of the sun! Over the horizon and across the plain the first shafts of the sun hit the hills. It may be the high crags of the precipice that receive the first flashes, but the splendor soon spreads broader and lower until the whole hill has its shining crown of gold. At noon the light is more diffused, more omnipresent, so that we do not perhaps notice its rays beating on the crest, but it is now in a greater glow than ever, notwithstanding no long shadows help out our imagination by forcing a contrast. At sunset the shadow of the hill is flung far along the plain to the east, a golden or pinkish or fire-red light returns to the crags; and the last light of evening, when all the plain is sunk in gloom, is that golden or reddish beam upon the rocky wall.

Catch points of sunlight.

Sunset on the crags.

Finally, there are the moonlight and the starlight of the hills with all their glamour; and on sultry summer evenings, when the twilight is nearly done and the air is hushed and no moon is in the sky, there is the distant flash of lightning along the crest. How the rivulet of light flashes upon ledge and dome, spreads over the

Moonlight in the hills.

forests, flares upon the heated air, and illumines the thickening sky! How huge, then, the mass of the hills, how much larger they seem in the dusk than in full sunlight! And what mystery in the vast gloom of the flash-lighted valleys, what romance in the golden-pinnacled tops!

The hills at twilight and dusk have always been great romance-makers with youth, great producers of retrospection with age. As we watch them the thought is about ourselves rather than the hills. The distant crags and domes merely put us in the mood. In themselves they have no romance and we strain the facts in speaking of

> "The vision of the hills and the souls of lonely places."

The hills have neither vision nor souls; they are merely inanimate rock or gravel, covered with grass or trees. But the romance produced in us by the hills is real enough sentiment. Put the cause of the emotion where we will, there is a poetry about lonely places and there are visions in the hills for those who have the eyes to see them. To youth the hushed loneliness and the far distance have to do with the romance of life and love, and the vision of the hills is a dip into the future. Always the thought is subjective. The lightning of the sky begets a lightning of the mind. We think fast and see futurity in flashes. We read our future in a mirror or a fountain, but we plan our action in lonely places. How many Davids have gone down from the hills in the morning into the camp of the Philistines! How much reality has followed fast upon the romance of the hills!

Their inspiration.

Aged eyes watch that flashing light, too; but it means to them something different, something deeper, perhaps. The vision is a hark backward to the long-vanished days of youth, and the quiet, the content, the deep peace of what remains. There is a realization of the goodness of life, of the wisdom of creation, of the beauty of the world. The hills and mountains were never "dismissed by nature from her care," but, with the lonely places, are the spots of earth where we can get back once more to nature's heart after a lifetime spent in the dreary Londons of the world. Back to the hills that sent us forth! The rain that goes down by the rivers to the sea and

Visions seen in the hills.

The return to the hills.

returns in clouds to the hills completes its circle not more truly than humanity. The Great Mother that led us out brings us back again to a resting-place in the quiet hills.

CHAPTER IV

FOOT-HILLS AND ROCK BASES

THE foot-hills are somewhat different from the worn-down mountains or eroded hills that perhaps surrounded the village of our boyhood and taught us our first lessons in mountain forms. As their name implies, they are the hills that lie at the foot of the mountains and usually they belong to a range or system. The word "foot" is not always, however, to be taken literally, because it implies closeness, whereas some of the foot-hills lie at a distance from the mountain bases. In fact, they may be separated by wide valleys and plains; they may be a little range in themselves and yet still belong to the greater range. However, the foot-hills of a range usually lie near the larger mountains and are regarded as stepping-stones to the higher peaks.

The original buckle and snap of the greater heights were probably repeated in the hills, but with declining emphasis. Thrust made them both, but the bend upward of the central peaks absorbed the thrust and left less

The out-lying hills.

Stepping-stones to higher peaks.

Less buck-led than the high peaks.

wrinkling for the lower hills. That the latter are less broken and splintered is not to be doubted; and when the arch has been snapped through and the rocks are exposed to view they are usually not the hardest, nor the oldest, nor the deepest of the rock strata. They are the last beds laid down in geological times and are sedimentary rocks of sandstone or lime- *Wrinkles of the outer crust.* stone. In other words, the wrinkling has apparently disturbed the outer layers of the crust more than the harder beds of granite lying deeper down. Occasionally the lower hills will expose granite or gneiss, but these are worn-down mountains rather than foot-hills.

Wash from the upper peaks. Everything wears down. The foot-hills bear the brunt of it, perhaps more than the upper peaks, because the accumulated wash from the peaks seems to be thrown off on the hills somewhat as an upper roof may shelve rain and snow upon a lower one. They are not only eroded by the ice and water gathered from their own exposed area, but they are gullied and undermined and slashed through by the swift glacier streams pouring down upon them and washing around them from the axial range. Any surface that lifts above the level of the plain is material for disintegration—food for the elements. In the flattening process the higher slowly settles upon the lower and the lower gradually pushes out into the plain and helps form the flat spaces.

The trend of the foot-hills is usually that of the main range—that is, they make lesser ranges paralleling the greater. If, as in the case of the Rockies, the trend is from south to north the foot-hills will generally correspond, even though they may lie off at quite a distance from the central ridge or continental divide. Occasionally there are cross-spurs that break through the line, making transverse valleys and mountain passes, but uniformity is usually peculiar to long systems like the Rockies. Not so with the Alps. The foot-hills there are as irregular as the main trend. The whole group shows buckling, twisting, wrenching in the most violent forms. Science can only conjecture as to how it was produced. But the result of central peaks and outlying foot-hills is there as elsewhere. Just so with the Carpathians or the Caucasus. They are flanked or headed or surrounded by foot-hills, according to the direction from which the original thrust came.

It has already been stated that the foothills are not the supports of the higher peaks, though the water wear upon them and the taluses at their bases create the feeling of a lean inward—a shoring up of the greater range.

Transverse ranges not buttresses.

Nor are the transverse ranges the buttresses of the main range, though they again create such a feeling. In the Pyrenees these cross-ranges run off from the main chain like the ribs from a spinal column. This is equally true of the California sierras. Often the main line is broken by long ribs leading off at right angles and almost always the buttress effect is

Feeling of support.

apparent. But the resemblances are superficial. There is no actual support, though there is a feeling of it which lends stability and strength to the whole range. Flanked or surrounded by outer walls, the central citadel seems unattainable, impregnable; and it is, perhaps, small wonder that people speak of

Eternal peaks.

the "eternal peaks." Counting by human centuries, we have never known time to wither or bend or bow their strength. They are as near "eternal" as anything within our ken.

Humanity in its settlements has never cared for the high mountains or their valleys. The mountain people, such as the Swiss, are exceptional and not too successful either physically or mentally. One feels always that they have pitched their tent in a harsh, rather ungrateful locality. But the races of men have ever been well disposed to snuggle up close to the foot-hills. The modern cities have been built beside great rivers on the flat lands, but in Biblical times cities were founded upon rocks in the lower hills; and where no hills existed, as in the Mesopotamian valley, the Assyrians built a huge brick platform in imitation of a flat hill and reared their city upon that. Athens was built upon and about the Acropolis, the Areopagus, and the limestone heights between Cephissus and Ilissus. Rome and Constantinople are both builded on their seven hills, and the hills around about Jerusalem are famous.

Mankind in the foot-hills.

Assyrian cities.

Higher and more characteristic foot-hills, however, lie back from many of the older cities. Out from Rome, a few miles, the so-called Alban Mountains appear as outliers of a higher range. In the ancient days the Romans built villas there, as at Tivoli, and, no doubt, rejoiced at the "view" across the campagna toward the Eternal City. What a view

The Alban Mountains near Rome.

it is even to this day! From Frascati, looking out at twilight, with the glamour of dusk lying upon the great plain, the twinkling of growing lights in the city, and a broad band of fiery splendor in the western sky, it is the very poetry of vision. And nothing could be more romantic than the same view a little later, when with the rising moon there is a purplish-blue effect spread through the air and the final flare of twilight gives the effect of Rome reddening in its ashes—the Rome of Nero burned to the socket.

The views from the Acropolis at Athens looking toward Phaleron Bay, from Constantinople over the Marmora, from Schönbrunn near Vienna, from the Catskill Mountain House, from the Poland Spring House are all so beautiful that we quite lose interest in the kind of hill we are standing upon. The Alban Mountains, for instance, are not foot-hills in the sense of being made by lateral pressure; they are volcanic upheaval. The crater lakes of Nemi and Albano suggest as much, and the tufa and the lava of the hills themselves confirm it. Truer foot-hills lie about Florence or

to the south of Bologna or at the north of Italy, where the railway from Milan to Venice passes along the southern base of the Alps. The rounded, well-wooded hills about Brescia or Bergamo or Verona are typical examples of Alpine outliers, as the mesa ranges of Sonora and Chihuahua are of the great Sierra Madre.

In beauty of form and color the foot-hills *Coast Range foot-hills.* lying along the California coast from San Francisco to San Diego are quite the equal of any in the world. They are made up of loose granite, are sparsely covered with short growths, and have been rounded by winds and rain until they are as smooth as the Scotch hills. In the afternoon light they take on hues of heliotrope as a body color and upon this are often blended olive shades of live-oak and sycamore—the whole flattened in a decorative pattern like tapestry. At Santa Barbara *Hills at Santa Barbara.* these hills come down to the blue waters of the Pacific with a strange mingling of warm and cool colors that is nevertheless harmonious and supremely beautiful. Farther south and back from the coast the bare, desert foot-hills of the San Bernardino range have only warmth of hue, but this warmth is wonderfully delicate in rose-reds, opals, lilacs, terra-cottas, and air blues. There is no mountain color like that seen in and about the desert ranges.

It is usually along the backs of the foot-hills that one's pathway lies in mountain-climbing. When they are in buttress form and join on the main sierra the ascent is very easy, especially if the ground is bare. You are then able to look back on the ascending steps of your climb, and nothing is more pleasing to the climber than studying his back-track. As one rises what wonderful views are revealed, what sweeps of valley depth, what lifts of mountain ridge! From the uplands of the Rockies in Montana, to the east of what is now the Glacier National Park, the look away over rounded foot-hills and sweeping table-lands to the great grass plains is a sight never to be forgotten. The wonderful beauty of the long hill lines, the great rest of the plateau lines, the sweep of the horizon lines make up a framing for a vast expanse of yellow grass dotted here and there by prairie lakes of re-flected cobalt blue. The wildness, the aloof-ness, the vastness of it are unsurpassed. The look backward from the foot-hills of the Alps over the plain of Lombardy is more pastoral, more civilized, more varied in color, but in-finitely less interesting in light, air, and color. The great spaces of the wilderness have a quality of beauty about them that no panorama of civilized lands can equal or even suggest.

As you come nearer to either the Alps or the Rockies, working your way through the valleys, you meet, even among the hills themselves, with the deep-cut ravine and the rocky canyon. They have usually been brought into existence by some glacier torrent or boulder-carrying stream that has chiselled and ground and worn its way downward into the rock until a great chasm is revealed. Sometimes the mountain defile is produced by a fault or break in the strata which has left a precipitous wall on one side and at its foot a small valley. In either case, by wear or by fault, a rock base is exposed and a portion of the crust of the earth is brought into view. The face of the wall is usually so sheer that the bare rock is almost the only thing seen, and its bulk and proportions are usually so stupendous that we are in the presence of the overpowering perhaps before we realize it.

No doubt the hardness, and hence the endurance and permanence, of the rock wall has much to do with its impressiveness. The whole earth is metal in an impure form, and a precipice is part of its hardened rim shown

Ravines and can-yons.

Rock bases.

Rock walls impressive.

Canyon of the Colorado.

to us by the chance cut-out of a valley or a canyon. It is a mere scratch in the surface, and yet when you are down at the bottom of the scratch, as in the valley of Lauterbrunnen or the Canyon of the Colorado, what an overwhelming impression it produces! The huge walls tell us that the world was not built merely for yesterday and to-day. It was con-

The self-supporting globe.

structed on the principle of a self-supporting globe—the strongest of all building principles —and though it may be liquid or only half-hardened matter within, yet its outer shell is of such massive masonry that it should last into eternity. It is braced by its own curve and supported by its own continuous circle. Nothing can crush it save contact with another flying world.

Precipices.

This feeling of bulk and power is intensified when the eye travels along such a face wall as that underlying the Jungfrau or struggles up the abrupt sides of the Rockies near Colorado Springs. The sides are precipitous, and yet they are hardly those "sheer precipices" which we read about in the tales of the trav-

Height of precipices.

ellers. In the low hills, as Mr. Ruskin tells us, one seldom meets with a wall from the top of which a plumb-line will swing clear for two hundred feet, and half the guide-book "abysses" of the Alps are slides of rock, or sloping taluses, or at best abrupt descents of only a few hundred feet.

In the high mountains, however, in places where the average tourist seldom goes, there are sharp declivities and veritable precipices that need no figures to make them awe-inspiring. On the Matterhorn, for instance, *Matterhorn precipices.* there are several places along the shaft where a falling stone as large as a water-bucket will disappear from view without striking the wall; and in many places in the Rockies there are look-overs into depths that must reach down several thousand feet—deep enough at any rate to make one dizzy even though lying flat on the rock and peering over the edge only with one's nose and eyes. Besides, there are *The Staubbach leap.* waterfalls like the Staubbach that actually do fall clear for nearly a thousand feet, and on the sides of Monte Rosa, in the Yosemite Valley, and elsewhere there are others of even greater sheer descent.

Perhaps our feeling about the overwhelming strength of the exposed rock wall is helped *Angle lines of the rock wall.* out by its acutely broken lines. The right angles are harsh, savage, even at times violent, and yet stimulating, elevating, exhilarating. Seldom do you see on the face wall that undulating line of beauty, that flowing continuity, which is found everywhere in the profiles of the low hills. The broken line prevails—the line that runs but a few yards or feet, perhaps, and then snaps off abruptly, to make the angle of another line. When one tries to reproduce *Drawing a face wall.* this appearance on paper with pencil or brush he meets with confusion. The lines run so at random and break without sequence that patience is exhausted before the scattered pattern is delineated. Even if caught the result is formless, chaotic, almost meaningless. But not so with the wall itself and the feeling it produces in us. It is a part of the great rim and has continuity, stability, permanence. *Suggestion of strength.* The zigzags and breaks of the face are in a measure indicative of its strength and endurance. They suggest that the terrestrial masonry is firmly set and massively braced against destruction.

And yet that broken surface illustrates and is a part of the old, old story of destruction. *Heat, frost, rain, and wind.* Its rock strata originally were flung haphazard there, with buckled and broken arches, perhaps, and now for centuries the gnawing teeth of the elements have been at work upon them. The breaks are produced by erosion —by frost prying outward on the ledges, by heat expanding and cracking, by rain washings and wind wearings. They all combine to loosen and crumble the walls of slate and granite, and one by one the fractured blocks drop down the face wall. The accumulation of fallen débris at the bottom builds up into a talus. After many years, perhaps, the talus reaches high up on the wall and becomes a huge litter of broken rock that turns the natural right angle of the mountain's foot into a concave quarter-circle. Behind the talus we can still feel the straight, descending line of the cliff—can still feel its prodigious strength and grandeur.

Growing taluses.

The grandeur and indefinite bulk of the cliff or precipice are often greatly promoted by its depth and spread of shadow. We seem to associate sheer walls with shadow if not with gloom. The sunshine falls upon them, to be sure, but we are left with no deep memory of it. Unconsciously we think of the upper peaks as being sunlit, but the precipice, with its "silent tarn below," is matter of half-light with its accompanying dampness and chill. The

Cliff shadow.

indefinite in the shadow produces the mysterious. The precipice becomes problematical and the imagination builds it in vaster proportions, in greater heights, in deeper depths, in more romantic light, in more poetic color than actually exist. Out of the shadow there come, perhaps, projecting points and ledges of rock that catch the sunlight and glitter by contrast, making of the face wall a wonderful pattern of light and dark. And perhaps, again, the wall itself is stained with strange local hues and saturated with indescribable tints and tones.

But color is usually no more thought of in connection with the walls of a precipice than is sunlight. We see where the drip of water makes streaks as black as ink or stains the wall in spots an orange-red with iron-rust; we notice the spread of lichens on the wall in large patches of gray, of yellow, of blue-black;

and, still again, perhaps we recognize beds of moss clinging along the ledges, wild flowers growing in the crannies, or even bushes or stunted cedars and piñons rooted in the cracks of the rock. But such color is usually patchy or at least not wide-spread.

It is, however, often very beautiful color

considered merely as an isolated patch. The rock flowers that grow in the ledges and hold fast with the mosses and thin grasses have a peculiar beauty of their own. They grow in the shadow, the winds toss them, the rains dash them, all the elements seem against them.

> "But patiently, slowly,
> The pale and the lowly
> Wring color and form from the stone."

It is hectic color—that fine tint of delicacy that comes to plants that starve—and the form is, again, of that sinuous, attenuated quality that goes with growths that are torn or worn by wind and water; but perhaps that very character of frailness and delicacy makes for beauty. By contrast with the rugged walls, the dark shadow, the drip of black ooze, the mystery of gloom, what lovely spots of light and color they seem! And, in spite of their pallor and their attenuated forms, how tenacious they are! Wherever there is a platform or shoulder or crevice in the rocks there they put forth a stem and blow a flower. They grow in beds along the ledges and gain a little, perhaps, each year. For ages and ages they live on, holding their numbers, keeping

Color of the flowers.

Tenacity of the growths.

Endurance of slight things.

their place, while boulders are falling away from the face wall and mountain tops are crumbling. Is this a survival through fitness? Or should the lesson read:

> "Not always the strongest
>   Endure and last longest,
>   Not always the fairest that fade,
>   The peaks and the towers
>   Wear down, but the flowers
>   Still bloom in the precipice shade."

Different-colored rock strata.

A larger coloring comes to the face wall when it is made up of different-colored rock strata. Frequently, in Colorado or Arizona or New Mexico, the entire mountain, precipice included, is of a red or terra-cotta hue from the oxidation of iron and copper. These colorings of orange, dull yellow, rose-red, silver-gray are things we often look at but do not see at all; and yet they are there. They are constant qualities of all mountain walls in arid regions. The table mountains, red bluffs, and

New Mexican escarpments.

abrupt escarpments of New Mexico and Arizona are remarkable for their layers of red, terra-cotta, and gray placed one upon another and all blended by a summer heat wherein the very air seems lilac-hued or opalescent. Similar hues are very apparent in the Colorado Canyon, while over in Algiers the English novelists gather them for "local color" and produce splashy effects with them that make the groundlings wonder. Of course, in much less degree, companion colors are spread upon the sharp walls of the Brévent or the huge bases of the Schwarze Mönch. The mountain walls of the Alps put forth no blaze of splendor—no poppy-field effect—but it is something of a mistake to think them merely a black or gray monotony.

Colors of Alpine walls.

Many of the precipices begin and end with a declivity, there being nothing above them but a bald top and a sloping back; but others support peaks and domes of snow that lift far above the base and sparkle in the sunlight. The Schwarze Mönch, of which mention has just been made, is the foundation wall upon which rises the Jungfrau. And from what enormous rock supports lift the Eiger and the Wetterhorn! Their ragged edges and steep declivities remind us again that they have not been geologically long upon the face of the earth. When they are many centuries older the tall peaks will have crumbled into round domes, the sharp ridges will have melted into smooth saddles, the precipices will have become

The Schwarze Mönch.

The Eiger and Wetterhorn.

sloping descents through the growth of their own taluses; grass and timber will, perhaps, cover the once barren rock.

Walls gnawed by the sea. Even in its youth the precipice and rock base are subject to sudden changes through earthquake-shock and the splitting or breaking of strata. Again, where a mountain range stands with its feet in the sea the face wall is almost always gnawed into scoops and hollows, and along the fissures of the rock strata the waves rush in and out wearing and washing long galleries into the heart of the mountain. *Ocean caves.* These are the beginnings oftentimes of the famous ocean caves where seals breed and mermaids play in the blue-green water, and the shipwrecked Don Juan clings to a scrap of rock and gradually wins his way back to life and love.

Similar results in rock-hollowing are sometimes obtained in the precipices back from the sea by wind and rain wearing in rock-fissures. *Wind-worn caves.* The desert mountains high up on the walls are often honeycombed with caves. But they have no great interest for the explorer of them. The interior of the mountain is not so romantic as the fairy-tale would make out. According to the tale the hero slips and is about to fall over the precipice, but saves himself by grasping a near-by bush. The bush tears loose just enough to reveal a stone step but not enough to let the hero fall over the edge. Of course, he explores the steps, finds a stairway leading into a great cave glittering with diamonds, rescues the captive princess, and, after many startling adventures in the depths of sapphire corridors and diamond halls, escapes to the open air again. *The cave of fiction.*

But the cave of reality is a trifle more prosaic. It is dark, not to say black, the air is usually suffocating, water drips from the rocks down one's neck, and the greater part of the locomotion is done on one's hands and knees. It is not quite so comfortable as a gallery in a copper-mine. Nor is it so interesting. Probably for these reasons it has not been tenanted since the days of the Stone-Age man. To-day not even a holiday party can be induced to tarry long in it. As a part of mountain beauty the cave need not be reckoned with just here. It is a defect—a blow-hole in the plate—rather than a depth of splendor. The beauty of the mountain lies on the outside. *The cave of reality.*

CHAPTER V

THE TIMBER-LINE

THE character or general look of a mountain chain may be produced by height, by the almost abrupt break of the axial range against the sky, by the peculiarity of peaks. The chain may be young and full of precipices or old and full of taluses; it may be snow-clad or bare-peaked or flanked by foot-hills; and, again, any one of these features would lend a determining character to the range. But, aside from the nature of the original upheaval, the external appearance of mountains is probably more influenced by timber than anything else. Rainfall—its presence or its absence—may make or unmake the look of the whole range.

For rain and, what is in effect the same thing, snow on a mountainside not only tear down and drag into the valley; they also build up barriers that check their own ravages. Rain produces the grasses and the mosses that cover the slopes and stop the wash of gravel and soil. It also grows the bushes and the forests that spread along mountain sides and hold back landslips. It further produces color, and this in itself is so distinctive a quality that often a range takes its name therefrom, as, for instance, the Green Mountains. Color is, of course, not the effect of timber solely. The Blue Ridge, the Black Hills, the Topaz Mountains take their names as the result of other effects. The blueness of ridges seen in far views is caused by the banked and accumulated atmosphere lying between us and the ridges:

> " 'Tis distance lends enchantment to the view
> And robes the mountain in its azure hue."

But it would not be so azure were it not for the green background of the forests. The "purple peak" of twilight is almost always a timbered peak. The desert mountains which have no forests have an opalescent atmosphere, and the peaks are not purple at twilight but often salmon-colored, rose-hued— even glowing red.

The southern ranges of the Rockies traverse an arid region and the timber growth is not thick or heavy. In places in the lower spurs

Mountains named from color.

Desert mountains again.

Chaparral and underbrush.

Coast Range forests.

there is only underbrush or chaparral. Farther to the north, as the rainfall is increased the timber also is increased; but even as high up as the Canadian Rockies, beyond Banff, the growth is small (though thick) upon the eastern slopes. On the contrary, when one goes over the divide toward the Pacific, passing through the Selkirks and entering a warm, damp belt, he meets with the huge firs of the Cascade Mountains and the tall, close-standing spruces, cedars, and pines of the Coast Range. Farther to the northwest, with increasing rain, the timber grows still denser.

Alaska timber.

In Alaska it not only wraps the mountain tops but extends down the sides to the very edge of the sea in a mantle of dark green. That mantle is practically impenetrable. The undergrowths combined with fallen timber make an axe necessary at almost every step. Even the streams that come down the mountain sides seem cut out of the timber rather than out of the rock. They flash here and there in spots, but frequently disappear and run under the great blanket of moss and decayed vegetation.

Alpine forests.

The Alps grow no such timber as that upon the Pacific slope, because they have no such warm climate, but the heavy rainfall of Switzerland brings forth somewhat similar if smaller forests almost everywhere. Possibly much of the present timber is of very recent growth, for it is not large; but it, nevertheless, covers the mountain sides and gives the distinctive blue-green coloring of the Alps. The characteristic look of Switzerland if reduced to a flat panel of color would be the green of timber below, the white of snow above, and the blue of the sky over all.

The colder the climate the scantier and the hardier the timber growth. Such mammoth *Effect of cold on timber.* trees as the redwoods of California will not grow up in the snow-line; and, while the Port Orford cedars stand thick on the lower slopes of the western mountains of Oregon, they do not thrive beyond, say, five thousand feet. It is the chill in the air that is responsible for the small, telegraph-pole larches of the Alps and the thin, schooner-mast pines of Norway. One need not fly from country to country to *Stunted growths.* find illustration for so obvious a fact; one need only ascend the mountain itself—Mount Rainier, for example—to see the timber dwindling in size and becoming more shaggy in bark and limbs as it nears the snow-line.

Inevitably the scantiness or plenitude of

Timber determining mountain character.

timber produces a distinct mountain character which the eye quickly recognizes. And each growth within itself develops a distinct character of its own. The larches and spruces of the Alps, fighting cold, wet, and tempest, clinging fast against water-wash and landslip, wringing existence from the thin, rocky soil, put forth abnormal features. The larches not only have thick barks but they secrete resins that keep heat in and cold out. They send out roots in every direction like the arms of an octopus, not only to pick up food and moisture, but to hold fast in the cracks of boulders and rock strata so the tree shall not be torn out by either tempest or landslip. The trunks, forced to grow out from the steep mountain side, bend and spring straight upward to the sun directly they emerge from the soil. And as though to forefend against wind they grow comparatively few branches. Again, they do not grow close together. Each tree is quite by itself and often there is open space about it for air and light.

Peculiar features of mountain timber.

Oregon and Washington timber.

Perhaps cultivation—the interference of man—has had something to do with the present condition of the Swiss timber, for it is a different story from the untouched tangle of trees and vines and underbrush that covers the western mountains in Oregon and Washington. There the trees stand so close together that they shut out the light of the sun and the undergrowth is often so dense that only a bear can break through it. As you look at one of these mountain sides from across a valley, so thick are the trees that the tops are hardly to be distinguished. They melt into a broom-like fringe of green—a bunched mass. The larches of the Alps, however, when seen across a valley, may be counted and checked off, each one with its arrow-head pointing upward at the sky. These tops even make a pattern that looks like a stretch of green tapestry, so regular and recurrent is the arrow-head design. Amost all of the Alpine slopes are thin in their timbering. You can, in spots, look through the trees to the moss and rocks beneath. And on almost every steep slope there are the bare troughs of descending slips and avalanches that have ploughed through the little forests as though they were mere corn-fields.

This Alpine slope is a delightful mountain side for the Alpine climber and his guides who are making an ascent. The timber does not worry them in the least. The chestnuts, walnuts, oaks, beeches that low down mingle with the conifers make a welcome shade; the spruces and pines are as picturesque as though they belonged in Benozzo's fresco in the Riccardi palace; and the larches are clean-stemmed, straight, and aspiring. Underfoot the grasses and the mosses grow thick, with occasionally some flower of the forest lifting its pretty head from beneath an enormous boulder; the lichens cling to every rock and tree trunk; and the beds of ferns, scattered here and there, offer no impediment to travel. In fact, mountain-climbing through the timber of the Alps is somewhat like a walk in a park, so open are the groves and so even the ascent.

Flowers in the timber.

But, again, for contrast, the pushing upward through the timber of Oregon or Washington or British Columbia is quite another affair. The trees stand thick, the underbrush is dense, the ground itself is uncertain and treacherous. At the foot of slopes where surface water settles there is usually a thick black ooze into which one may sink to his waist; in other places there are some miles, perhaps, of dead and fallen timber, blown down by hurricanes or half burned down by fires. These "windfalls" and "burns" are sometimes criss-

Growths on Pacific slope.

Windfalls and burns.

crossed by three or four layers of fallen trees until they begin to look like an enormous bear-trap. It is impossible to get through them with a horse, and the traveller on foot may proceed only by walking along the fallen trunks high in air. Ancient windfalls where the timber is, perhaps, embedded in the ground and overgrown with moss are quite as difficult to get over, for now one keeps sinking through the trunks that yield to the foot like so much ashes. To step on a giant fir that has been mouldering undisturbed for several centuries is often to go down into it up to one's neck. *Decayed firs and pines.*

Nor is the travelling smoother in other places. Great ferns and a huge thorny weed called "devil's club" reach up to one's head, bushes of yellow and red salmon-berry, blackberry, raspberry, thimbleberry tear at one's clothes, wild grape and poison-oak tangle about one's feet. By way of compensation and distraction one sees overhead, say fifteen feet in height, vast canopies of rhododendrons bearing clusters of flowers a foot or more in length, and far above these the spreading foliage of innumerable pines, firs, and incense cedars. *Difficulties of travel.*

The stand of this giant timber (some of the firs and spruces are six and eight feet in

Giant firs and spruces.

diameter and perhaps three hundred feet in height) is astonishing. It is the forest primeval, hoary with long beardings of caribou moss —the original forest produced by nature at her leisure and in her own way, uninfluenced by man. The magnificent prodigality of the growth has no equal on the face of the earth.

Thickness of the stand.

It would seem in places impossible to put in another tree, so thickly do they grow, and yet the young trees are springing up everywhere. The beauty of the individual trees is matter of common observation. The straight stems lift and taper so magnificently, spread branches so picturesquely, group tops against the sky so wonderfully, that the dullest of tourists

Dead firs.

would exclaim over them. Even the giant fir with bark stripped off, that has been standing dead for years, has a rhythmical rope-twist in the trunk that reminds one of a baroque Italian column and a silver color that is a marvel of delicacy.

The redwood.

But the most magnificent of all the giant trees, in its columned trunk, red bark, and light-green foliage, is the *sequoia*, or redwood. It does not belong high up in the mountains, but flourishes low down in the California parks, and yet is more or less of a mountain tree. Its height is often over three hundred feet and its diameter over thirty feet. The bark alone is two feet thick and the rings of the largest specimens make them out some four thousand years old. You cannot see the tree sway with the wind, and it makes no moaning as does the pine. The young ones have pointed tops, but the older ones are ball-topped or ragged and scarred by lightning. These huge tops are vivid green and reflect splendidly the first light of the rising sun, the last of the setting sun. Next in height to the *sequoias* are the Douglas spruces, the sugar-pines and yellow pines—all of them superb growths with tapering stems straight as arrows and with arrow-headed tops.

In its initial stage the tendency of all this mountain timber is to spindle upward toward the zenith, to throw the foliage into the top, and to let the trunk go bare of limbs. The young conifers (lodge-pole pines and mountain hemlocks especially) with bare stems and bunched tops are very tall for their bulk and sway much. Even when there is little movement of air they sway majestically and slowly, and bow significantly to one another in the gravest manner. In a wind they rock a great deal from the ground up. When older grown they bulk heavy in the trunk, become stiffer, and stand stiller. In time all the conifers—pines, spruces, firs—become massed and packed in enormous areas and spread over hundreds

of miles. Into these great woods the sunlight comes only in broken flickers, the rain does not drive but drips through the covering boughs, and the winds creep down only in gentle puffs of cool air. Tremendously impressive are these huge forests with their shadows and their stillness—too much so for the average traveller, who usually prefers the open, sunlit places.

And open places do occur even in the Oregon and Washington woods. There are spots where the sunlight falls free and a varied vegetation springs up. Occasionally a stream comes down a mountain side, splashing around huge boulders and tumbling over shelving rocks into pools where rainbow trout lie hidden. Along the banks of the stream, growing out of the very rocks, as it were, are willows, poplars, birches, live-oaks, chinquapins, ma-

droñas. Among them are white tangles of shadbush and syringa, mauve-colored rhododendrons, with many varieties of laurel and azalea. They seem to gather thickly by the edge of the stream, but whether to supply insects for the trout or misery for the fly-casting fisherman is not very apparent. They seem quite unconscious of either fish or fisherman, blossoming there in the open spots of the wilderness as they have done for many years, alone and undisturbed.

The forest is usually not the place for flowers. There is too much shadow with its consequent damp and cold. The long lists of Alpine flora that we find in the journals of Swiss tourists are made up from the meadows and the uplands rather than the woods. The heavy timber region is practically flowerless. But this is not true of the forests of Oregon and Washington. Nature there is so productive through warmth and moisture that she will grow stamens and petals even in spots where the circumstances are not altogether favorable. The wild rose, that dearly loves the hedgerow in the open, will often cast a profusion of pink petals against the red bark of an incense cedar, and the large, purple flowers of the clematis sometimes cling about the sides of a giant fir in a wild blare of color. What they do there, how they came there, no one can more than

Flowers of the forest.

Clematis and roses.

guess. Nature's ways are always a little mysterious. Why should she put sunflowers in the forest—out of the sun where they naturally belong? Yet they are to be found in abundance in the Siskiyou woods, with tiger-lilies, paint-brushes, red and yellow columbines, lupins, and many varieties of orchids—all of them apparently hardy, productive, resplendent, and yet all of them quite out of place.

If one looks among the grasses and mosses for the low-growing flowers the variety is increased. Anemones, spring beauties, triliums, baby blue-eyes, forget-me-nots, violets are massed in beds and scattered at random through the forest. It seems as though anything would grow in that wonderful soil, in that wonderful climate. Even the weeping spruce—the rarest of all trees—is found there, and there only. Every inch of the ground sends forth something rare and strange. Sometimes in crossing a dense windfall, following along a tree trail perhaps fifteen feet in air, one may see the ground beneath him thickly strewn with little drooping, bell-shaped flowers lifting no higher than the awns of the moss, or again great patches of star-shaped clusters making a yellow carpet of fairy beauty. Again,

Orchids in the Siskiyou woods.

Small flowers.

The weeping spruce.

what do they there in the wilderness? If nature made the beauty of the earth for man alone, why were these fair creations born to blush unseen? Is it, perhaps, possible that she grew the flowers in this primeval forest that their beauty might *not* be seen by man—might *not* be trampled upon and turned into ashes? The edelweiss once grew everywhere on Alpine slopes, but now you will find it only in the most inaccessible spots. Have the flowers the same instinct of self-preservation as the birds and the animals, or is it only the remainder, the portion that has escaped destruction, that we now find in the remote corners of the earth?

The birds usually use the woods as a place of refuge. In times of storm or summer heat many varieties in pairs and flocks may be found hiding in the cover, but, with the exception of the woodpeckers, the nuthatches, and the jays, there are few that breed or live in the woods. In the spring of the year the European birds are to be found on the plains of Lombardy or the fields of France or Bavaria—anywhere on the flat lands about the Alps rather than in the high forests of the Alps themselves. In the lower timber there is

Petals that blush unseen.

Instinct of plants.

The wood birds.

some bird life, but even there it is abnormal or peculiar. That is to say, the birds are often small, drab-colored, wag-tailed, ill-balanced, but very agile, very swift in movement. They bear some odd resemblance to deep-sea fishes that live in darkness. The majority of them are insect-catchers, finding food under the bark of trees or in decayed spots of the trunks. It is astonishing at what distance they can see a minute fly and how accurately they can dash at it and snap it up. Some of them are abnormal in voice, having a note much too loud and deep for their size. Many of these war-blers—chits, wrens, vireos—build nests in the holes of trees or cracks of rocks and keep dashing in and out like mice from a hole in the floor.

In the Pacific-slope woodlands there are a few more birds than in the Alps—a crow, a purple jay, a large black woodpecker, a black grouse, a mountain-quail, some bluebirds, fly-catchers, rock-wrens, and snowbirds, with a water-ousel that bobs up and down from a rock in the stream. None of them sings or gives forth a joyous note. The jay jangles and squawks, limb by limb, from the bottom of a pine-tree to its top, then flies to the next tree and repeats the performance; but he does not sing. And the woodpecker pounds holes in the trees and then pounds acorns into the holes; but he does not sing. The calling note of the quail or the grouse and the whistle of the bluebird are sporadic utterances only, and the note of the water-ousel is rarely heard for the roaring of the water. Farther south, in the California mountains, where the timber is thinner and there are many open parks, there is a corresponding increase in bird life, but the dense forests are usually songless.

There are animals in the deep woods, too; but, again, they seldom make any cry. They seem living there for protection—living there because there is no living elsewhere. Time was when bear and deer were frequently seen in the Engadine valley, but they have passed out and become merely a tradition. The chamois and the marmot still exist, but not in the woods. They keep to the barrens and the rocks, and from their shyness are evidently aware of their danger. It is only a question of time when they will go. Switzerland is too well known and supports too many hunters with magazine rifles for animal life to thrive.

White-tailed and mule deer.

Such is happily not yet the case with our own mountains, though they are threatened. They are being tramped over and hunted by bands of alleged sportsmen, but they still have their quota of game. The white-tailed deer, or the mule-deer, is in almost every Western range between Mexico and Canada. The mule-deer often prefers the chaparral to the woods, whereas the white-tail likes the timber or river brush. Each seeks the cover best suited for his hiding, but eventually the hunter catches him napping and the soft-nosed bullet ploughs into him. Year by year their respective numbers decrease, and at no distant day they may follow the buffalo to oblivion.

The grizzly.

At home in the chaparral.

The same fate is in store for the grizzly bear, only it will come swifter and surer. Every man's hand is against him. Guns and poisons and traps are always confronting him, and, notwithstanding his enormous strength and courage, he is doomed to extermination— is even now almost unknown. He knows the white man is on his trail, and to-day he skulks in the densest chaparral of the Coast Range, where neither dog nor man can follow him with safety. When brought to bay and wounded he is probably the most dangerous of all big

game. A blow of his paw will crush the skull *Strength of the grizzly.*
of a man like an egg-shell, and he can snap
the neck of an ox instantly with a single
muzzle twist of that same paw. Fortunately
he is not often met with. The chaparral is
hardly good hunting-ground, and no sports-
man, unless unnecessarily ignorant, is likely to
invade it. The thick, dry brush catches and
holds the man as the web the fly, but the grizzly with his enormous weight smashes through
it as though it were so much standing corn.
That weight is variously estimated at from a
thousand to fifteen hundred pounds. And yet *His endurance.*
there is no want of swiftness, agility, or endurance about the beast. He will stand shooting through the heart and not collapse until
some minutes after he has torn the enemy
into ribbons. Decidedly he is an animal to be
let alone.

It is persistently asserted by Pacific-coast *The Rocky Mountain silvertip.*
naturalists that the real grizzly lives (or did
live) only in the Coast Range, and that the
Rocky Mountain grizzly is another species
quite different in weight, color, and habits.
It is denied that the silvertip is a grizzly.
The silvertip is close of kin to the big brute
that has become so tame in the Yellowstone

National Park that he eats from the garbage heaps in the rear of the tourist hotels. Nor are the Alaskan bears—huge as they are—true grizzlies in the Californian meaning of the term. The typical old Ephraim belongs in the chaparral of the San Bernardino mountains or elsewhere in the neighboring ranges. But his numbers are very few to-day, and the day of his departure is near if not already arrived.

Everything gives way before the coming of the white man. Beasts and birds, with trees, bushes, and flowers, all disappear before him. He is the great destroyer. There are to-day in the western mountains bears, wolves, cougars, lynxes, wildcats, deer, porcupines, coons, marmots, mountain-beavers, squirrels, rock-rabbits, but none of them will last for long. Perhaps they will not be greatly missed, for they are now unseen by the average person in the forest. They keep out of the way and seldom give out a cry.

The mountain forest is, indeed, a place of overwhelming silences. Even a storm there is something that goes on overhead but is not felt down among the tree-trunks. What a stillness, that of the big woods! In the summer there is sometimes a hum from the great Æolian harp of the pines overhead, or in the winter the cracking of a frost-bound branch or the almost soundless sifting of snow from the upper branches—no more. The jangle of a jay or the dull thump from the bounding hoof of a deer is but a momentary affair. Usually the stillness is so great that it may be felt, the silence so oppressive that it may be heard.

And at night with the silences comes the feeling of being lost, being shut in, being utterly helpless until the coming of light again. Alone in the forest at night—a camp-fire sending flickering flashes up the red trunks of the surrounding cedars, the stars seen only in occasional points of light through the arabesque of boughs, a great unknown of darkness all about you, and occasionally out of that darkness the glittering yellow eye-balls of some beast watching you from behind a fallen spruce—is picturesque, undoubtedly, but not exactly comfortable. One welcomes the dawn. The first touch of the sun that strikes across a fire-burn and gilds the tops of the distant firs may seem more beautiful than anything ever seen on land or sea, and the jangle of the jay may sound really sweet after a night with the forest silences. Even the Great Peace may at times prove oppressive.

Serenity of nature.

And yet how serene nature is in her mountain stillness, her forest hush! How aloof from impatient fret stand the tall firs, the pines, the cedars! The snows fall and melt and pass away, the streams slip down from the mountain to the lake, the laurel and the columbine nod drowsily along the banks, the water-ousel on whirring wing goes up and down the brook. How utterly indifferent they are to everything but their own endeavors!

Nature building up and tearing down.

Nature works not by time but by circumstance. A millennium is as a day to her. She builds up and she tears down; she brings to life and she brings to death; but always for a use and for a purpose. And always with beauty as a result of the work. Even in the death there is a final touch of beauty. The white-trunked and ghost-like pines blasted by fire years ago—even the dead lodge-pole pines still standing and covering the lower mountain-tops by the million—will reflect the most beautiful tones of rose and lavender at sunset.

Again, all the foliage is at its most brilliant pitch just before it falls. And the monarchs of the forest, when they themselves succumb, what lovely mosses and flowers gather upon the decaying trunks!

Death is a part of the plan, in the forest as elsewhere. It is not or should not be cause for regret. Hence, perhaps, the feeling of serenity, of mighty repose, of supreme and abiding purpose when we are close to the great elemental forces. The fret of the world seems purely human and artificial. Nature in her larger manifestations is above and beyond it.

CHAPTER VI

THE UPLANDS

WHERE the timber ceases to grow there the uplands begin. By uplands I mean the semi-barren lands lying between the timber and the snow—the mountain sides where boulders are thickly strewn and small streams wander and stunted bushes with short grass and hardy flowers grow. Occasionally there are patches of dirty snow in the deep-sunk gullies or ravines, but usually the uplands are pale green with short herbage or pale gray with scattered rock taluses. In still summer days the tourist from his mountain hotel may occasionally hear the faint tinkle of cow-bells coming down to him from those upper meadows—meadows that are so far up he cannot see the cows that wear the bells.

And here, too, with the ending of the forest is the beginning of mountain-climbing. In Switzerland, as already suggested, the tramp up through the timber is usually a leisurely matter, with many stops by the way. If near some popular resort the walks are duly graded and seats are placed at opportune eminences, where one may rest and look out upon a "view." But the canny *kurverein* can rarely be induced to build pathways up into the snow-line. The tourist must hire a guide to show him where to walk, unless he have sufficient experience and can pick his own way and do his own climbing. It is usually not difficult walking on the barrens, though there are sometimes precipices to be skirted and landslips to be avoided, but as one ascends the air becomes thinner, and one is increasingly likely to find himself "fat and scant of breath." *Difficulties of the climb.*

It seems that in order to gain your second wind you must lose your first. The process of breathing freer and fuller, which is called "getting one's second wind," is perhaps simple enough in explanation. It takes half an hour or more of climbing to get the lungs expanded to their full capacity and the heart beating with regularity. When this is attained one is in good walking condition and moves along without labored breathing or fatigue. This is needful, for sometimes the ascents of the uplands make stiff climbing, and the winds that blow up there are not zephyrs from Arcady. Scrambling up the warm lee side of a *Getting one's breath.*

mountain and emerging upon the crest to face an icy wind is exhilarating but also somewhat dangerous to one who is overheated or exhausted.

But what an air it is that blows on the uplands! It seems to come down out of upper space, so keen it feels, so clean its touch, so free from earthly taint. And so refreshingly cool! Is there anything more life-giving than the high mountain air in your face as you wander leisurely along an upland crest or pause for a moment on some moss-covered boulder! Small wonder that physicians sometimes send their patients six and seven thousand feet up on the mountain side. That air has proved many times a veritable call-back from the grave.

Nature is not always so cruel as she seems. She is merely exacting. Live as she designed you to live and you may live long. It was not of her planning that you should live in a city, in the air of a hothouse.

Nor did she plan for our spending most of our lives in the shade. The sunlight is as essential to the human as to the tree or the flower. And where, again, shall you see it or feel it in such purity and splendor as high up on the mountain! Coming through a dust-free air, it gives small hint of dispersion into color. It is not yellow or rosy but comes very near being pure white light. The great sun-shafts that break through loopholes in the clouds are like bursts of silver straight from heaven; and where the beam falls upon the snow of an upper peak what a dazzling spot of white it produces! One gets back to the pristine purity of the elements as he rises on the mountain side. *White light.*

And with the purity of the air, the clearness of the light, what vistas one sees from the uplands! What pictures are framed in mountain heights and valley depths! From Piz dell' Ova Cotschna (8,890 feet) above St. Moritz, for instance, there is the look to the northeast down the valley of the Inn with its shelving slopes and grass-grown meadows rolling like folds of green velvet—a peaceful view very beautiful in the afternoon when bars of sun and shade are thrown across it, but not a startling showing of upper peaks and sharp-edged precipices. On the contrary, the view in the opposite direction toward Chiavenna is astonishing in its tumbled and tossed heights and its canyon depths. A deep defile leads to the southwest with an array *Views from the uplands.* *Above St. Moritz.* *Toward Chiavenna.*

of lofty peaks on either side of it. Snow and glacier, mountain wall and waving waterfall, clouds and sunbursts, with snow caps seen through the mist, are all there. Down in the valley are the dark surfaces of two peacock-green lakes, and far beyond Maloja, where the Alps dip down to Como, the gentian blue of the sky fits into the notches of the mountains like a block of old Byzantine mosaic.

From the Bernina Pass.

At the top of the Bernina Pass, looking to the south, the view is even more impressive, though quieter and less distracting to the eye. Scores of mountains are pitched upon a sloping field—sloping downward again into Italy. Mountains rounded, pointed, flat-topped, and square mingle with mountains gray, blue, reddish, and clay-colored. Aspiring turrets belted with heavy courses of precipitous rock look like enchanted castles of the Titans, and misshapen domes of whitish limestone heave into the blue with a startling color effect.

From the Kölner-hütte.

A very different panorama—another tourist view—stretches before one from the Kölner-hütte (7,626 feet) near Karersee in the Dolomite region. There you see a hemicycle of white mountains with the Ortler group in the centre. The white peaks begin at the extreme left with the Fleimsthal, the Latemar, the Brenta, the Adamello. Then come the Ortler, and swinging round to the right the Otzthal, the Stubai, and the spurs of the Rosengarten. At your back are the walls of the Rothenwandt. There are four ranges—four planes in the panorama that the eye meets and crosses—before you reach the Ortlers. The blue air lying in drifts in the valleys rises with the heat of the sun. The distant peaks show faint and far. They are snow-covered and above them small clouds with flat bases hover like distant aeroplanes. The huge valleys are stretches of dark green, dull blue, and pale purple; the rocks and precipices are gray-yellow, orange-stained with lichens, blackish with dripping water, terra-cotta-hued in out-croppings of strata; the sky-line is a ragged edge of snowy peaks; the sky above is like lapis lazuli. The true mountain-lover looks at such a panorama perhaps for a long time but says little. The vast array of ranges is too big for praise, or poetry, or comment of any kind. It is something we accept without com-prehension—something that stifles the imagi-nation and leaves us, perhaps, with a dumb wonder.

The views to be seen from the uplands in America in some respects are even grander than those in the Alps. The look across the plains to the Mississippi, which may be had almost anywhere along the front range of the Rockies, is a marvel in its inclusion of the large elements. Earth, air, and sky are enormous spaces made sublime by extent and mass and not broken or weakened in force by disturbing details. This is quite as true of the Colorado Desert seen looking east from the San Bernardino peaks in southern California. A huge desert world reduced to its primitive elements is before you. There is nothing but range upon range of bare porphyry mountains and valley after valley of desert sand and gravel; but what massive strength in the bare rock, what majestic grace in the drifted sands! And, above all, what a superb opalescence in the air, what a radiant splendor in the sky! It is an enchanted land, soundless and tenantless, full of mystery, uncanny in its coloring, desolate to the last degree, and yet perhaps the most enthralling, fascinating landscape under the sun.

Quite different from the desert views are the sights to be seen from the high benches of the Mexican mountains below the City of Mexico, looking down the western valleys to the Pacific. The valleys are robed in tropical green and the thread-like lines of many streams flash in the sun. Towns, towers, white walls of ancient *haciendas*, white gleaming roads are there. Fields of rice and cane with groves of orange and rows of palm check the surface at times. It is a beautiful, even a grand, sight. But we seldom concern ourselves with that phase of the view. Our wonder is caught and held by the enormous expanse of the valleys—twenty-five miles in width by a hundred or more in length—and we feel only the world sweep of their far-reaching lines. After that Mexican view down to the Pacific all the rest of the world seems petty or confined or over-crowded with small things. Nothing goes beyond it save the great reach of the Himalayas.

Western valleys of Mexico.

The sweep to the Pacific.

When the wonder over the world panoramas begins to pall we turn once more to the smaller beauties on the mountain slope. The green and yellow uplands that lie back of St. Moritz-Dorf are interesting to walk over aside from the air, the light, and the view. At first one marvels, perhaps, over the appearance and dis-

Upland meadows.

Larch and juniper.

appearance of growths.   The larches go out at six or seven thousand feet, but why, one asks, does the dwarf juniper come in at that height?

Alpine flowers.

Daisies and buttercups flourish in the meadows down at sea-level, but the Alpine climber will also find them at ten thousand feet growing on the snow-line.   They are the same daisies and buttercups, only smaller in stem and lowlier in stature.   And they blossom and seed at the same time as those down in the valley.   Violets, orange-colored dandelions, bluebells, clovers, mountain pinks do substantially the same thing; but why do the Alpine rose, the pink heather, the edelweiss insist upon living on the wind-swept uplands and not elsewhere?   Every one recognizes, to be sure, the general statement of fact that

> "From their nature will the tannen grow
> Loftiest on loftiest and least sheltered rocks
> Rooted in barrenness,"

Hardy character.

but why and wherefore the nature that despises warmth and nourishment, preferring a starved and lonely life on the uplands?

The provision of nature is always more or less surprising.   In desert sand scorched with heat or mountain rock frozen with cold, it might be thought she would not attempt to grow plants—that she would leave the spot bare to sun and wind. But no. Just where she is expected to fail in resource there she puts forth a shrub or grass or flower that fights the elements for life, clings stubbornly to its barren bed, and brings forth a progeny as hardy as itself to carry on the struggle.

Nature's resources.

Of course it is a dwarf growth.

The dwarf growth.

"In that bleak upland, that rare mountain
   height,
Where sun and frost succeed by day and night,
What other than the dwarf could come to
   light?"

And frequently the growth is wiped out entirely by the prevalence of conditions necessary to its very existence.

"The warmth that stirs the juniper to spread
   Its clinging arms along a granite bed
   Unslips an avalanche above its head."

But in spite of circumstance and accident the plants live on. The Alpine growths are all of them good strugglers. They fight not only the long, cold winter but the night chill of summer. It is almost always frosty at night

*The strug-
gle for life.*

Plant devices.

on the uplands; clouds gather early in the day, and snow may fall at any time during the summer. Besides, there are winds that whip and tear at the lowly growths as though they would wrench them from their rootings. The moss lies flat and presents small surface to the wind, the gray and orange lichens cling tenaciously to the rock and move not, the larch and the arolla (the Swiss stone-pine) have their needles that foil the gusts, besides cold-repelling resins; but what protection have the flowers—the gentian, the harebell, the primrose, the cowslip, the columbine, the narcissus, the wild geranium, the azalea? They would seem to be at the mercy of the elements and the animals, and the only way nature can keep them from extinction is by breed—by prodigality of numbers.

Prodigality of flower growths.

But here nature does not fail. There seems no such thing as limit of numbers to the growths. The daisies and buttercups lie as thick on Swiss slopes as in American meadows; the lupins, dandelions, yellow pansies, violets, forget-me-nots, gentians grow in beds that sometimes spot the hillside for miles; and many a hanging valley is yellow-hued with great fields of waving broom. Very beautiful are these flowers in their clear, pure coloring, their freshness, their mountain quality. If they were refined in color or threw off much perfume they would hardly be appropriate to their blue-green environment. Again, if they were very delicate they would not stand the severity of the climate. It is their Alpine character *Their Alpine character.* that we admire. Beauty may lie in character and fitness quite as effectively as in form or color. And there, again, nature never fails. Her growths are always marked by adjustment to need and place. She grows no garden roses on the Alpine snow-line nor does she plant edelweiss on the plains of Lombardy.

The wild animals of the Swiss barrens are *Swiss fauna.* few where the flowers are many. They have largely disappeared. Occasionally, in protected regions, the Alpine climber sees a little bunch of chamois—sees them disappear over a ridge— but it is an unusual sight. The chamois is a *The chamois.* short-bodied, long-necked antelope with the habits of a goat. He lives on the uplands near the snow-line and browses on the short grass and bitter plants that grow there. His brown coat blends into the coloring of the rocks and herbage so closely and he stands so still that you do not see him, but he has not

His equipment.

the slightest difficulty in seeing you. He is always on the alert, always suspicious. If you watch him while he is grazing you will see him nibble at something for a moment, then stop, look all around, go to the edge of a precipice and look down or stand with head on one side looking up. If he sees nothing he will nibble again and then repeat the inspection. He is well equipped with eyes, nose, and ears, like all of the antelope family, and each sense is very acute. Perhaps his eye is the most wonderful feature of his make-up. It protrudes from its socket somewhat like a star-sapphire from a ring. It sees in all directions and renders the animal impossible of approach except from behind cover.

Agility of the chamois.

Strange and varying stories are told of the chamois's agility and jumping powers. Some writers make them jump anything up to fourteen feet high and eighteen feet wide. As for going down mountain sides or over precipices, the tales told are still more fantastic. They certainly have high powers in jumping crevasses, and, being goat-footed, they do some amazing jumping along the ledges of cliffs; but they hardly go over abrupt precipices, staying or guiding their fall by dragging their hoofs along the sheer side, though they are sometimes seen humped up in straitened circumstances on a precipitous ledge, and one wonders how they got up there or how they will ever get down again. Strange animals, driven up from the meadows, up through the timber-line, by man, they have adapted themselves to the barrens and the snows and cling there as tenaciously as the stunted herbage upon which they feed.

On mountain ledges.

> "And still they live, these children of the rock,
>   Gray moss, dark shrub, the meagre chamois
>   flock,
>   Their natures trained and tempered to the
>   shock.

> "Nor live in vain, the hardy mountain clan
>   Are but exemplars of creation's plan,
>   That all shall fight for life and those shall live
>   who can."

The only other animal of the upper regions, besides the chamois, that one sees is the marmot. Even he appears much oftener in advertisements for summer hotels than in reality. He is only a stupid woodchuck that spends most of the year sleeping in a burrow and the rest of the time dozing on a sunny rock. The

The marmot.

The whistler.

Swiss specimen is about as big as a rabbit, is grayish brown, and in many ways is not unlike the American prairie-dog. A larger specimen is met with above the timber-line in the northern Rockies and is known by its whistle. The older hunters and trappers still call him the whistler. He has a dirty white neck and shoulders and is quite as lacking in energy as his Alpine cousin.

The uplands of the Rockies are in many ways different from those of the Alps. In places of rainfall the grass and short herbage are thicker and of a brighter green up to the very snow-banks. As for the floral display, it is quite baffling to describe because of its great variety. Almost everything that grows in the Alps and hundreds of varieties besides are to be found. As the altitude increases the growths become hardier and grayer until finally they disappear. But while disappearing, so far as our eyes are concerned, some of them may be growing under the snow with a strange insistence that their time for flowering has come and that the snow is holding them back. It is astonishing in the summer-time what chill and frost they will endure, survive, and even conquer. The avalanche lily

Rocky Mountain flora.

(sometimes called adder's-tongue) not only grows under the ice-sheet that hardens along the edge of the snow-line, but it grows *through* a foot or more of ice and comes out to the sunlight. The hole through the ice that it melts or drills is not more than half an inch in diameter—just enough to let the flower stem pass through. Thousands of these stems may be seen pushing through the ice-sheet, and where the ice has receded and left them free thousands more are in bloom with delicate pale petals trembling in the breeze.

In June, as the ice-sheet or snow-line recedes, flowers of all kinds follow it. Spring beauties, paint-brushes, buttercups, lavender anemones come up out of the cold ground in hosts innumerable. A little behind them, a little later, come the purple lupins, the red and purple pentstemons, the phloxes, the squaw grass, the asters sometimes mistaken for daisies. Close up to the snow live also such hardy growths as the purple and white heather—a true-enough heather but different from the Scotch variety. Indeed, the upland slopes of the Rockies from Mexico to Canada grow peculiar plants that seem to have only a general likeness to plants elsewhere on the globe. And all of them are

Brilliant coloring.

excellent fighters for life. The cold and the frost add a zest and result merely in a heightened color. For it seems as though the color of the plants here was even more brilliant, more primary, than with the Alpine flora. Of course, all the mountain growths have learned their lesson well throughout the ages. They all tell a similar tale of simplicity, hardihood, fortitude.

Rocky Mountain animals.

The animals of the Rockies that are found above the timber-line are said to be arctic in character because the same species are found within the arctic circle. Preferring cold to warmth, it seems immaterial to them whether they live upon mountains ten thousand feet up or live in valleys farther north and nearer the pole, for in either place similar temperature and practically similar food are obtainable. The moose, the elk, the mule-deer, the bears do not go above the timber, as a rule, and are not distinctly upland fauna; but the sheep and goats are peculiar to the region—that is, they are seldom found elsewhere.

The bighorn.

The bighorn is the type of the Rocky Mountain sheep. There are several varieties found in different parts of the mountains from the desert ranges of Arizona to the snow ranges of

Alaska. They are all remarkable for the immense size and nearly spiral curve of their horns. The tales told of the wonderful things done with those horns—the downward leaps over precipices, the turning over in air, the alighting uninjured on the horns—quite rival the stories about the Swiss chamois. In reality, the only use the bighorn has for his horns is in fighting his own kind. He is a most agile climber and quite as well equipped with protecting senses as the chamois. He is always on the lookout from inaccessible ledges or high ridges where he wanders with a sure foot. He likes to graze on the uplands at night and morning, going up higher in the mountains during the day. As an adaptation to harsh surroundings he is a great success, and it is something of a pity that his numbers are so rapidly dwindling. In the inaccessible heights of the southern Rockies, even in the desert mountains of southern California, he still exists in small bands, but gradually his kind seems to be falling farther back into the ranges of British Columbia and Alaska.

Habits of the big-horn.

The same tale of disappearance is true of the Rocky Mountain goat. He is now rarely found south of British Columbia and is much

Rocky Mountain goat.

scarcer than the sheep. Perhaps that is because he is rather stupid and does not keep so sharp a lookout as he might. He goes in bands, following certain beaten trails along the rocks and across taluses, and his undoing may be put down to these habits. The only safeguard for health and long life that he possesses is a white, hairy coat that is not easily seen against a snow background. In addition he chooses for his habitat the very loftiest ridges, where, being sure of foot, he travels with safety and ease. He is a great climber but not at all remarkable as a runner. Nor does he jump over precipices and alight upon his horns. As a jumper he has some local reputation, but it is not that of the bighorn.

As for the birds, one is not able to discover many varieties above the timber-line either in Europe or America—that is, birds that live there during the season. In the Swiss barrens there is a thrush-like bird that nests along the edge of broken banks in the heather or moss, some snowbirds, some sailing hawks and owls, but no song-birds save an occasional sparrow. In the Rockies it is practically the same tale. There are eagles, buzzards, and owls, and on the southern Coast Range the condor is still seen occasionally. In isolated spots at the north one finds a white grouse or ptarmigan that belongs, perhaps, to the arctic circle. He lives on the uplands much as the grouse upon the Scotch moors, save that he is a hardier bird, has a rougher environment, and is a more beautiful and perfect creation. *The white grouse.*

The birds, animals, and plants of the uplands are all alike in that nature tempers no winds to them, smooths no pathways for them, has no watchful eye for them. All that she does is to increase the tap-root, the plumage, the muscling, the senses, the intelligence. They fight their own fights and struggle with their own struggles. That out of the strife comes strength, mastery, high development in one case, and failure, death, extinction in another case is apparently matter of no concern. She furnishes the equipment and the environment. For the rest, let each growth look to itself. Stern law! But how inevitably right! How otherwise than from adversity shall be wrung those dominant and abiding qualities, fitness and character, wherewith the world sustains itself and is a thing of beauty forever! *Nature's equipment.* *Laws of life.*

CHAPTER VII

MOUNTAIN WATERS

THE uplands are not only the first lands to catch the drainage from the upper snows, but they are also in the belt of heavy rain-precipitation and are the first to feel the descending showers. As a result they are usually wet lands having much boulder and gravel, much thin grass, moss, and lichen, many small streams, and numerous pools and lakelets. The glacier streams lower down are also made up from the drainage of the snows, but they are usually much larger, they gather and run in deep valleys, and have quite another look and character.

The little pool is something found almost everywhere in the high mountains, but it seems more appropriate, more at home, on the grass-and-rock barrens than elsewhere. It is usually a shallow mirror, bright in reflection of sky, cloud, and mountain peak, but nothing more. If it is the catch-basin of rains and its bordering shores are heather-covered or moss-grown, it may have a pale-ale hue about its water; but ordinarily it will be clear and quite colorless. That is, it will have no local hue of its own but may reflect the color of anything about it and be influenced somewhat by the light or dark of its bottom. Should the pool or lakelet be formed at the base of some small glacier its water will be jade-colored or Nile green by reason of the granite and mica particles suspended in the water. A large lake produced in this way, where much of the sediment has a chance to settle at the bottom, will be a peacock green in color with streaks of lapis-lazuli blue, or perhaps variations of blue and green dependent upon the light. Neither of these brilliant colors improves the mirror quality of the lake, for the best reflection comes from the clearest water.

In clearness of reflection nothing could be more perfect than Crater Lake at the head of the Rogue River in Oregon. It is seven thousand feet above sea-level and sunk a thousand feet in a round crater that is seven miles in diameter. It has no visible intake or outlet and is supposed to be the direct result of snow and rain. Lying deep down, it is usually smooth or only slightly ruffled by wind. And its reflection of sky, clouds, and encircling walls is astonishing.   The reflection is a trifle darker than the object reflected, and the hue is a very delicate blue-green.   The lake is something more than a guide-book wonder; it is a marvel of light and color.   The Swiss or Italian lakes are not comparable to it.   They are much less refined in tone and far coarser in the reflection.   Yet the dark Swiss lakes with their intense blue-green have a splendid, jewel-like coloring of their own which makes up for any slurring or blurring of the reflection.   That Walter Pater spoke of them as "horrid pots of blue paint," and shut his eyes in crossing the Alps that he might not see them, argues nothing against their beauty but something against over-refinement of the æsthetic sense. It is possible to refine one's sensibilities to the point of attenuation.

Almost all of the smaller mountain lakes— and they grow smaller as the altitude increases until they become known, poetically or romantically, as "lakes of the sky" and "mirrors of the cloud"—have outlets in some little stream that goes dashing down a mountain side to join a mountain brook.   At first the brook is a very modest little stream, has shallow banks, winds under huge rocks, stops in little pools, or falls gracefully over some ledge of rock into a basin where bubbles come and go and small trout lie hid and the sunlight flickers on the falling water. Almost always these little brooks, at some point in their winding, will cross an open mountain meadow where daisies and buttercups are standing in clumps, and columbines, perhaps, are swinging and swaying over the hurrying water, and yellow-and-blue beds of low-lying, nameless flowers are woven in the thick green grass. Very lovely and pure are these brooksides with their flowers. They are nature's precious tapestries, and not even the dazzling snows and sunset hues of the high peaks above them can dull their color or dwarf their delicate beauty. *Across mountain meadows.*

After passing the meadows the brook may become larger and more noisy, plunging and leaping over huge boulders down steps and terraces, running swiftly under majestic firs, flashing brightly in spots of sunlight—a glittering thread of life on the dark mountain side, a voice in the silent forest. What enchantment in the murmur of that water! The waves of the seashore will beat into your brain a tragic monotone—a dirge or funeral march— *Mountain streams.*

but the mountain stream gurgles, babbles, and tinkles with suggestions of fairy bells in Elfland. It is joyous, fanciful, entrancing—even under the midnight stars when you are lying beside it and cannot sleep for the loneliness of the wilderness about you.

The growing brook.

As the brook winds down still farther it grows by the accretions of side streams, becomes fuller, and, where there is gravel or loose rock, begins to cut a deep channel in the mountain side. But perhaps the flat rocks of the underlying beds soon force it to the surface again and spill it over some high ledge into the air. Then we have the mountain waterfall, with which every traveller is familiar.

The waterfall.

Many of the high mountains have illustrations of the mountain stream being shunted over a precipice—the waterfall shot out into the air. Sometimes it is a permanent feature of the landscape, but more often it is seen only after a heavy rainfall. In the Alps an hour's downpour is sufficient to make all the mountain sides run streams and waterfalls in lines of foaming silver. Of course they dry up and disappear a few hours after the rain ceases, but while they last they are graceful and brilliant affairs, falling, as many of them

Alpine waterfalls.

do, several hundred feet and often breaking into spray and mist before the ground is reached.

Of the mountain waterfalls that dash and wave the whole year through, the Staubbach in Switzerland and the Yosemite in California are, perhaps, the best known to the average traveller. The Staubbach sways with the wind, pitches down *[The Staubbach.]*

> "in a glittering flight
> Like a torrent of stars from the Bowl of Night,"

and, being of only moderate volume, it dissipates into water-dust before reaching the valley. The Yosemite has a larger stream and falls some twenty-six hundred feet in three leaps, again swaying in the wind and scattering clouds of rainbow spray on either side of it. *[The Yosemite falls.]*

> "As it sinks and breaks
>   Into cloud and mist,
>   The water-dust takes
>   Hues of amethyst,
> And across it thrown, in a gleam and glow,
> Are the spectrum hues of the bended bow."

There are waterfalls in every well-watered mountain district though few of them fall so far

Cataracts not in mountains.

as the instances cited. As for the huge cataracts like Niagara and Victoria Falls, their downward pitch is slight compared with the smaller streams, Niagara being a fall of only one hundred and sixty feet and Victoria four hundred feet. It is the volume of water pouring over them that makes them impressive. Such a volume of water would not be possible on a mountain side. Even if a sufficient source of supply could be obtained so high up in the mountains, the force of the falling water would mean swift destruction to both the mountain side and the valley. All the great falls are in the valleys and belong to broad rivers.

The re-united waters.

After a mountain torrent has dashed itself to pieces down a precipitous height its fragments usually come together again at the bottom; and, presently, the reunited stream wears for itself a new channel and moves on as before. As it descends it continues to grow in size, cuts a deeper bed because there is more earth and soft débris into which it may be cut, and, as it carries in its water sand and small gravel, its power of cutting, even into

Wearing a canyon.

hard layers of rock, is increased. The tendency of it is to wear back into the mountain wall, and, after many centuries of this wear, perhaps a gorge or canyon comes into existence.

The word "canyon" always brings up the thought of the great slash in the ground caused by the Colorado River, just as the word "cataract" immediately suggests Niagara; but I mean neither of these, for neither of them belongs properly to the mountains. There are a hundred steep defiles in the Pacific Coast Range that have been made by torrents and through which torrents still pour on their way down to the plain and the sea. They are all somewhat similar, and perhaps the one I have now in mind will answer in description for the others.

Pacific Coast canyons.

The canyon of the Coquille River in Oregon is down in the timber-line and has abrupt walls of rock with mosses and lichens clinging to them, with flowers and ferns and stunted bushes growing along the ledges, with surface-water oozing from the niches and falling in an endless drip on the rocks below. Down in its bed of boulders the little stream boils and pitches and roars, cutting and wearing out a deeper bed for itself as the years go on. What it has already done is apparent not only from the high, water-worn walls on either side

Coquille River canyon.

but from the rounded boulders in the stream, from the shelving shores of rock, and from the enormous pot-holes that have been worn in the rocky shores. Some of the holes are like huge cisterns, are a dozen feet deep, eight feet in diameter, and almost as perfect in their rotundity as though a machine had cut them. But the whirling water with a hard, round stone for a grinder was the only machine that worked upon them.

There are falls in this Coquille River with rock basins, running rapids, babbling shallows, and deep pools lying under overarching trees. It is more of a brook than a river and has all the characteristics of the mountain stream. The gorge it has cut is narrow, even dark in places—so dark that the sky overhead seems like a mere ribbon of blue stretched across the top. Heavy precipitation of rain or a bare mountain side where the water runs off with swiftness, carrying with it huge stones that fracture and cut their way, seems neces-

sary for the production of such a cleft. The desert mountains of Arizona or the Californian mountains supply such conditions even better than those of Oregon. The upper Rogue River in Oregon, King's River or the Kaweah in

California, are picturesque and perhaps more familiar examples than the Coquille.

There is nothing quite like these Pacific canyons in the Alleghanies or the Pyrenees or the Alps. The pitch down from Maloja to Chiavenna is now only a steep, descending valley, the road up to Zermatt follows a roaring torrent but has no enclosing walls, and the precipices of the Lauterbrunnen valley probably never formed a canyon at any time. The valley coming down from Karersee to Botzen is decidedly more of a canyon effect with its upright walls of rock, and is profoundly picturesque; but the valley between Predazzo and San Martino is more characteristic of the eastern Alps with its sloping meadows of grass, its patches of pine, and its flowing stream that turns the mills, irrigates the meadows, and gladdens the life of the valley dwellers. *Alpine gorges.* *Dolomite valleys.*

These canyon streams are rather lonely places, and one may watch the play of water over rocks sometimes for hours without interruption from man, beast, or bird. And yet there is some life there. Somewhere along almost every one of them will be found a pair of water-ousels. They are said to be shy *Loneliness of the canyon.*

birds, but perhaps that is merely a way of
accounting for their being found in lonely
places. They fly with a rapid-beating wing,
alight on stones in the streams, and bob up
and down for some seconds before they begin
to move about. Then they run in the shallow
water, dive, push their heads under the water,
and turn over the small stones looking for
food, much as the starling uses his beak to
turn over dead grass. They are very active
in wading and diving and are even said to
fly under water, but this latter feat is seldom

observed. Besides the ousel there are a number of fluttering, long-tailed little birds that
live along the streams and catch flies, bugs,
and beetles for a living. They are expert in
manoeuvring in the air for a June-bug or a
beetle, and generally succeed in breaking the
back of the quarry at the first snap. In repose they all have trouble in keeping a proper
equilibrium between head and tail and teeter
a good deal. The more common mountain
streams running across meadows and through
open timber, such as that coming down from
Paradise Valley on Mount Rainier, for instance, have, of course, a greater variety of
bird and animal life along their banks.

Glacier streams in the different mountain ranges have some similarity of character one to the other and yet are not exactly alike. And they differ from what we have called the canyon and the mountain stream even in the same range. The latter are frequently the direct distillation of the clouds caught by a surface drainage and carried quickly to the lower lakes or rivers; whereas the former is the seepage from the bottom of a glacier and is quite another story. The glacier stream does not flow so easily or so smoothly as the mountain stream. It pours over a boulder bed and is retarded by boulder shores, but it has an enormous push about it and its oscillations from side to side have a savage cut. Doubtless from the sand and gravel carried in the water it is heavier than the mountain stream, which may have something to do with its powerful swinging movement.

A mountain stream sometimes runs so rapidly over a boulder bed that the surface of the water is thrown into ridges and furrows. The tops of these ridges, as though unable to keep up with the pace of the under water, have a way of curling back, apparently plunging up-stream like leaping salmon that fall and are swept down again by the running water. All shallow streams that run swiftly show these ridges and hollows, but the glacier stream, perhaps, shows them less than the others, and where they appear they are heavier in movement as though the water itself were heavier and less pliable under pressure. One has the same feeling about the rapids, waterfalls, and cataracts of glacier rivers. The falls of the Bow River at Banff are tremendous in volume, but they seem to fall with unusual heaviness. One wonders if the sand and gravel in the water have anything to do with the sullen thundering plunge. The Kicking Horse River, full of grit and fry, gives the same general impression.

Bow River at Banff.

If there is any question about the weight of the glacier stream, there can be none whatever about its color. Everybody recognizes that color at once because it is peculiar to the snow-mountain streams. It is not a green, a blue, or a brown, but a milky or sand color when the stream is heavily charged with silt, and the color of gray-green jade when the water is deeper and freer of sediment. This latter is the local color of the glacier stream, and has little to do with sky reflection because

Color of glacier stream.

the jade color shows as readily under clouds as under blue sky. The color is influenced somewhat by the gravel bed over which the stream runs, but the particles of sand and mica that are in the water itself form the main coloring matter.

As the stream grows in volume and winds down the open valley at a slower pace, it begins to clear though still retaining its jade hue. The waters that flow from the Bernina glaciers change in tone when they reach Samaden, and after they have entered the Inn clear still more, but not until some miles have been run. The Rhone at Geneva, even after it has had a chance to deposit its sands in Lake Geneva, passes out of the lake with a greenish glacier quality about its color. The infinitely small particles of sand and mica are still in it and give it that tinge of local color which has been so variously reported upon by travellers.

The lakes that form in the mountain valleys—their beds sometimes scooped out by glaciers and their waters supplied by glacier streams—exhibit many phases of blue-green coloring dependent upon the quality of sedimental deposits combined with reflections from

*Emerald
Lake near
Field.*

sky and forest and mountain side. Emerald Lake, near Field in the Canadian Rockies, is brilliant in peacock blue and green, less, perhaps, from its solutions than from its reflections. It is deep sunk in the hills, surrounded by heavily timbered mountains, and between the reflections of blue sky and green trees it is almost impossible to get at its local color. Besides, there is an unusual blue air in the valleys about Field. That makes a dark-blue sky which in turn spreads a darker reflection upon the lake. Lake Louise, more directly fed by glacier streams, and more in the open, is less intense in coloring though still a decided blue-green. *Iceberg Lake.* Iceberg Lake, a small sheet of water on the eastern slope of the Rockies, in what is now called the Glacier National Park, has, perhaps, a more exquisite tone of blue than either of the Canadian lakes. It is really a glacier pool at the bottom of an amphitheatre of rock—a *cirque*, so called—and is fed by small, hanging glaciers. The water catches sky and rock reflection from above and its local hue is again difficult to determine, but it is, perhaps, nearer an air blue than any other tone.

The angle of reflection at which a lake is seen, of course, has much to do with its appearance. When you are several thousand feet above the lakes in the Engadine and look down upon them they show the weird blues and greens of Böcklin's paintings. One portion of the surface rippled by the wind will, perhaps, show green; another portion in the shadow of the mountain side or darkened by the late afternoon shadow of trees will be blue-black; and alongshore, where a surface stream swollen by a recent rain is pouring sandy, muddy water into the lake, there is a decided jade color or Nile green. Again, Lake Nemi, in the Alban Mountains, is a crater lake, having nothing to do with glaciers; yet when seen from the top of Monte Cavo, looking down upon it, it shows again Böcklin's blues and greens. The steep sides of the lake—some three or four hundred feet or more in height—down which pour the little drainage streams, are responsible for the green, as the sky overhead for the blue. Corot, who painted so many pictures with the title of "Lake Nemi," never gave the slightest hint of these colors. He always showed the white-light reflection which came from his painting down close by the water and looking across it at a flattened angle of vision.

That is, if he painted by the lake at all. The steep shores, the rocky and wooded banks, the heights of Monte Cavo, the Italian sky, are as absent in the Corot pictures as the blue-green water. He was never too true, topographically, to any place, and was very fond of painting a Corotesque landscape in his studio and giving the result an historical or classical or geographical name afterward. His "Ville d'Avray" is quite as false to local fact as his "Lake Nemi," but his convention was not the less beautiful and, from an art point of view, perfectly justifiable. Art is not a map-making affair.

The proximity of Monte Cavo to the lake suggests another genius who somewhat tortured local truth just here to produce an effect. Monte Cavo is the Alban Mount of *Childe Harold*, and it was from its height that the ocean was duly Byronized in Spenserian verse.

The famous apostrophe was another studio production—written first, perhaps, and located afterward. The sea from the top of Monte Cavo does not "roll on" to any appreciable extent. In fact, it is not seen except on clear days, and then only as a band of blue beneath the sky. Besides, it is not the ocean but the

Mediterranean, which is usually in rather a docile mood just here. However, there was the romantic association of near-by Rome which justified Byron's distortion, as Corot's; and in any event poetry and painting are far removed from topography.

As far, perhaps, as brilliant prose-writing may sometimes be from dull truth or probability. Ruskin has some compelling sentences (which I am far from belittling) in his *Modern Painters* to the effect that it is the designed mission of the mountains and the hills to feed the plains with living streams, thus promoting the fertility of the earth, the purity of the air and sunlight, and tending to gladden the existence of man. In the mouth of the great preacher that is not a bad conceit, but in the reality of nature it is not so very apparent. In British Columbia, and sometimes on the east coast of Mexico, the mountains pour down such superabundant streams that the land is inundated and man is incontinently drowned out. In the Andes all the rain descends and practically all the streams run on the eastern slope, so that Bolivia is an uninhabitable jungle, while Peru, on the western slope, is a desert where man is ever on the dead-line

Ruskin on mountain waters.

of thirst. All of which suggests that the mountains on the western continent sometimes shirk their mission on earth and raise doubts in prosaic minds about their having a mission at all. Perhaps they just "happened"; and perhaps, again, they have poured their waters through the ages down upon the plains just because water would not run up-hill.

CHAPTER VIII

GLACIERS AND AVALANCHES

THE streams that roar down the mountain sides, leaping over precipices and breaking into clouds of spray, do not carry away all the snows that gather on the peaks. Frequently the upper snow-fields, by thawing and freezing, become packed into a bed of ice called a glacier, which, perhaps, moves down a ravine not more than a foot or so a day. The glacier in the course of years finally reaches into a lower valley, where it melts and passes off in a drainage stream to the lowlands and the sea. It seems a connecting link between the peaks and the valley, for it starts in the cold and ice up above and sometimes ends down below in a summer meadow of trees and flowers.

But the glacier is not seen creeping down every mountain valley. In the temperate zone only the high, snow-clad peaks send them forth, and even among these certain conditions are necessary to their formation and flow. An adequate snowfall, a thaw-and-freeze temperature, an altitude or a latitude, a suitable bed are primary factors. The Carpathians, for instance, are not high enough for glaciers, while the Andes have an insufficient precipitation with few slopes or mountain plateaus to hold and congeal the snowfall. The Himalayas are higher in the air by a third than the Alps and have many snow-fields on their tops, with a few hundred-mile glaciers; and yet the Alps are, perhaps, more famous for their glaciers and offer better illustrations of the glacier *Alpine glaciers.* type. The Aletsch Glacier is not the largest and the longest in the world, and the Grindelwald Glacier and the Mer de Glace are tourist-trodden and almost commonplace, but they are excellent specimens of the ice-river—better than anything in the central Rockies or anything accessible in the Caucasus.

Glaciers in the Rockies. The comparatively light snowfall on the Rockies and the Sierra Madre accounts in measure for the scarcity of glaciers there. Farther to the north, in Canada, in the Selkirks, and in the axial range of the Canadian Rockies, there is an increase of glaciers with the greater snowfall, and when one reaches the region of heavy precipitation in Alaska the glaciers become enormous. In these high latitudes the ice-river does not always melt and resolve itself into a stream some thousands of feet above the sea, but frequently comes down intact to the very sea itself, snaps off on its submerged end, and the broken part floats out to sea as an iceberg. Still farther north, in Greenland, this is the ordinary way in which glaciers run their course. The whole ice-cap of Greenland has something of the character of the glacier and from year to year keeps slipping slowly toward the sea.

The Greenland ice-cap.

The formation of a glacier usually begins with the fallen snow on the slopes or ravines or plateaus or amphitheatres of the mountains. Whatever the character of the snow when it first falls—whether flaky or crystal-shaped or spiculed—it soon turns into *névé*, or granular snow. As it settles down it packs, and by packing, thawing, and freezing it forms into a porous snow-ice at the bottom. This ice is almost always crystalline in aggregation—that is, made up of many-sided crystals. It may be broken or crushed by squeezing, or changed in look according to the circumstances of its formation, but the crystal is its component factor. The disposition of the crystals in glacier ice is not structural but haphazard, and quite different from that in lake ice, which

The névé.

Crystals in glacier ice.

is formed in columnar masses up and down. Moreover, the snows of successive seasons are stratified in the glacier ice and lie in layers one upon another, so that one can tell the depth of each season's fall and, were the glacier exposed in cross-section, the number of snowfalls it contains.

The first slip downward.

In its first formation on a high mountain-top the tendency of a snow or ice bed is to spread and slip from the centre downward. A small ravine or hollow receiving the downward push from the slopes forms a snow or ice drift which, in turn, soon starts and moves at a foot-a-day pace down into the valley.

The ice tongue.

This is the ordinary *valley* glacier. It is merely an ice tongue proceeding from the central cap, and would be not unlike an icicle if the icicle would lie flat and slip instead of hanging down straight. It is the most common form of small glacier and is sent down by almost every high range of mountains.

Piedmont glaciers.

When several glaciers starting from a common snow-field reach into separate little valleys and come together at the bottom, or when one huge glacier deploys on the mountain side and the deploying parts afterward come together again down in the valley, the result is called a *piedmont* glacier. Sometimes a deploying branch, or even the main ice-stream, arrives in a lower valley in a more violent and hurried manner by being pushed over a precipice and gradually breaking off on its overhanging end. It keeps falling at intervals in broken fragments almost like a frozen waterfall, not, however, with the roar of waters but with the thundering shock of the avalanche, and amid clouds, not of spray, but of ice-dust. It is now called a *hanging* glacier and is starred in the guide-book as something not to be missed by the visitor. This kind of glacier is not common, neither is it infrequent. The Victoria Glacier seen from Lake Louise in the Canadian Rockies shows a familiar example.

All these forms of glaciers actually move in their valley beds, like rivers of ice, but, of course, very slowly and methodically. It was thought for a long time that they were motionless—waters frozen in the valleys—but there have been many proofs of their movement both romantic and scientific. For instance, in the year 1820, three guides were lost in a crevasse high up on the Mer de Glace. Their bodies were not recovered until 1861—

forty-one years after—when they came out with the ice at the foot of the glacier. That was proof to the lay mind more positive, if more gruesome, than any scientist's figures. Nevertheless, the scientists have tested it and proved motion in other ways quite as convincing.

Practical proof of movement.

A row of stakes driven in the ice across the Mer de Glace from shore to shore revealed the fact that the stakes were moving very slowly past fixed points on the shores. The figures given for it are about one hundred and fifty or two hundred yards a year, or, as Tyndall estimated it at different points along the glacier, from twenty to thirty-six inches a day. The movement of all glaciers is not, of course, so slow as this. The condition of the ice, whether wet or dry, deep or shallow, gorged or spread out, taken in connection with the temperature of the locality, the roughness or smoothness of the glacier bed, the inclination or slope downward, are the factors determining the speed. Thus, while the flattened Mer de Glace moves slowly, and requires a hundred years for the snows that fall on the Col du Géant to arrive at the end of the glacier, the Muir Glacier in Alaska moves at the quick-

The Muir Glacier.

ened rate of seven feet a day, and some of the glaciers in Greenland during the summer months exceed fifty feet a day.

There are other ways in which the glacier, though an ice-river and apparently rigid, resembles the water river. For instance, the row of stakes across the Mer de Glace after a few days showed a bend down-stream, indicating that the middle of the glacier moved faster that the shore sides of it. Another test showed that it moved faster on the top than on the bottom and faster where gorged than where spread out. Like water, the ice-stream is retarded by friction and advanced by volume. It is sometimes higher in the middle than on the sides, and, though it is a thin, sluggish mass, brittle in parts, it still submits to squeezing and bending, and even oscillates from side to side somewhat like the running river.

Once more, the glacier registers a heavy fall of snow almost as accurately as the river a flood of rain. Down at its valley ending it varies in length from year to year. Last year perhaps it advanced far down into the valley because of a heavy snowfall in the mountains, say, a hundred years ago. This year perhaps

Middle and top movements of glaciers.

Oscillation.

Registering of snow-falls.

Crevasses.

it will retreat far up the valley because the snowfall of the following year was a light one.

The surface of a glacier is covered with crevasses, or cracks in the ice, of varying width and depth, and these are caused by the friction of motion. Along the shore edges marginal crevasses, extending outward into the glacier, are produced by the shore masses not being able to keep up with the more central masses. A split or crack takes place at right angles to the stress. The midstream end of the crevasse moves faster than the shore end, until, after many weeks perhaps, the whole crevasse has turned about and is now almost parallel to the shore instead of at right angles to it. In the meantime new crevasses extending out from the shore cut across the crevasses already made and now at oblique angles. This process keeps repeating itself until finally the whole shore edge of the glacier looks like a piece of crackled glass.

Crackling of the ice.

Squeezing and crushing.

Then there are the crevasses produced by the top moving faster than the bottom, by squeezing and crushing, and by the inequalities of the bed. When a water-river goes over an abrupt ledge of rock it breaks into a rapid, but when an ice-river goes over the same kind of "bumper" it splits and cracks across its top, opens in seams, but still holds together at the bottom. So that, all told, there are crevasses of many kinds—marginal, radiating, longitudinal, parallel—that crisscross and cut through each other in a very confusing way.

Opening of the crevasses.

The tendency of all these huge gaps in the ice is to become wider and deeper as they move down farther into the valley. Besides, they are opened wider by the melting of their sides, by rains, by dripping waters that pour down them. Almost every glacier gathers upon its back more or less débris in the course of its many years of slow descent into the valley.

Glacier gathering of débris.

The chief accumulation is dirt, sand, gravel, and boulders that drop on the glacier's back from the overhanging rocks of the shore or are rolled down from the neighboring slopes. These are dark objects upon which the sun's rays centre, and the result is a melting of the ice about each rock and a sinking or pitting of the rock itself in the ice. The drainage from these various pits runs to the low part of the glacier's back and forms into little lakes on the ice—lakes not deep but sometimes over a hundred feet long, as that, for instance, of the Victoria Glacier. As reflectors of the

Glacier lakes.

white peaks of the neighboring mountains and of the blue sky they are astonishing, while in quality of blue color they are the most exquisite of all the mountain mirrors.

Each one of these miniature lakes (often seen with little icebergs in them) has a streamlet that drains it, and the streamlet runs but a very short distance before it plunges headlong down a crevasse. This makes what is *The well or moulin.* called a well, or, from its churning sound, a *moulin.* For a time, perhaps, the bottom of the well remains firm and holds the water— the most beautiful well with its blue depth and its blue walls of ice that ever snow nymph or Alpine climber looked into—but gradually the falling water wears through the bottom-ice and flows away with a gurgle under the glacier and out below in the valley.

The accumulation of rocks and gravel on the back of the glacier, especially on the portions along the shores, is gradually carried down by the movement of ice, and as the ice melts they are deposited in ridges parallel *Lateral moraines.* to the movement of the glacier. These are called *lateral* moraines, and eventually, by accumulation, they make banks or sides for the glacier. Sometimes a glacier by its great weight and pushing power digs up the gravel and rock in its own bed and pushes the mass ahead of it into the valley. This is called a *frontal* moraine. Frequently the drainage stream coming out at the foot of the glacier carries with it enormous accumulations of dirt, silt, sand, gravel. This glacier wash extends perhaps for miles down the valley and is known as the *valley train*. Sometimes the hole, or tunnel through the ice where the stream flows, gets clogged with its own débris and the deposit finally appears as a long ridge of gravel and small boulders. This is called an *esker*. These various accumulations are seen about the glaciers of almost all the lofty peaks, but, again, possibly the Alps show them better than any other mountains—at least people see them there as they do not elsewhere.

When the glacier ends in the valley the ice, through rottenness, may lose much of its beautiful blue color, though where a stream comes out at the bottom of the glacier and above it is an arch of ice the blue hue may still hold strong. The Yoho Glacier in the Canadian Rockies is a fine illustration of this, with its three-hundred-foot span of azure vault above

Frontal moraines.

Valley train and esker.

The Yoho Glacier.

and its jade-hued stream below. Several Swiss glaciers have tunnels cut in their ends by escaping streams, or dug out by man, into which the tourist is taken by his guide to see the blue of the ice—a blue quite as marvellous in its way as that of the famous grotto at Capri. But none of these tunnels is so impressive as the Yoho arch.

The color of the ice, the fantastic bridges occasionally made by melting, and the wells of water held in the ice are about the only attractive features of the glacier. Its movements and its peculiar formations are curious scientifically, but æsthetically they are hardly inspiring. To this last statement, however, there are exceptions. In Alaska, two miles or more to the south of the Taku Glacier, is an enormous, flat-backed, dirty glacier that moves down to the sea, slowly crumbling and melting on its front like a huge sheet of rotten granite. It is covered with gravel and boulders and has no form or color to attract one. It is interesting only on account of its size. On the contrary, the Taku Glacier near it has size and something more. It is some seventy-five miles long, and on its face wall where it breaks into the sea it must be three-quarters of a mile in breadth and probably several hundred feet in height. If the Palisades of the Hudson, where they break abruptly to the river, were a little smaller and made up of ice they would look not unlike the front of this glacier. The up-and-down structure of the Palisades rock is similar to the ice formation of the glacier save that the latter is finer, more flaky, more crystalline.

Imposing and beautiful as this ice structure is in form, the beauty of the coloring goes far beyond all else. One never sees, in any of the mountain glaciers of the temperate zone, such wonderful blues and greens as those in the Alaskan glaciers. The hue is exquisite, jewel-like, and in depth unsurpassed. Even when the ice breaks off in huge blocks and slowly floats away as a small iceberg the splendid color goes with it and is reflected in the water. The Alaskan glaciers (they are only the southern tongues of the great arctic ice-cap) are not mere matters of wonder, they are things of beauty. The great ones, like the Muir, the Pacific, and the Valdez Glaciers— the latter with fifteen miles of frontal ice-cliffs—have an even greater beauty of color than those more accessible.

Alaskan glaciers.

Great glaciers.

All glaciers are usually seen at their worst by the average traveller. By that I mean they are seen only in the summer-time and at their disintegrating valley end. High up in the mountain ravine the glacier—snow, ice, and water—is much cleaner, purer, more lovely in color. It is there simply a frozen mountain stream and, like the stream, is pure at its source however muddy and polluted it may be at its exit in the valley. Grindelwald Glacier and the Mer de Glace low down are, largely, so much rotten ice mixed with stones and dirt being slowly pushed into a valley. It is a different tale higher up. And different, too, in the winter months. Then the stream running away at the bottom of the glacier is shrunken in size but clear, the glacier itself is muffled with snow, the snow-fields are glittering white, and perhaps the only sound heard is the faint murmur of water under the snow.

In intensely cold weather even the water-murmur ceases, there is contraction and congealing, with little crackling sounds occasionally running through the glacier mass. The glitter of the snowflakes increases the feeling of the cold. For scores of years, perhaps, the water has been lying there locked up in snow and ice crystals. The little crackling sounds seem almost like threads of pain or cries of souls in bondage to be free. A fanciful soul might think them calling:

> " The cold, only the cold,
>   Are garbed in white and gray;
> The cold, only the cold,
>   Shrink from themselves away.
> Lord of the sun, shine on us!
> Warm us with heat and rain,
> Turn us back into water,
> Make us a river again!"

The slowness of the glacier moving down a mountain valley finds its striking opposite in the swiftness of the avalanche. The snow that slips and slides from the roof of a house is a poor miniature likeness of the iceslip of the mountains. The Alpine avalanche descends with a sound and fury that signify danger and destruction. Again, the glissade of light snow that a chamois's hoof may start, and that slips with a hissing sound down some steep slope, may knock a mountain-climber's feet from under him and carry him over a precipice; but usually it works little havoc among the trees and meadows. It goes down in a

The avalanche.

The snow glissade.

cloud of snow-dust and is soon over with, even if it is not dissipated before it meets with the pines. But the avalanche of ice, the huge snowslip, and the landslip are very different affairs.

Following well-worn slides.

Avalanches that follow well-worn paths, slipping down narrow ravines with much noise and enormous puffs of snow, again do little harm. They are the accumulations of ice and snow formed in the shallow catch-basins of the mountains which finally become overloaded and overweighted. That which serves to start the mass downward is sun, wind, and rain. These causes result in melting, which in turn loosens the grip of the mass on the mountain side, and presently the whole sheet is slipping down the ravine with not a tree in the way to check it. In the summer

Falling avalanches.

months the sound of these falling avalanches can be heard every hour, and the hours of the early morning—the moments of the greatest night cold—are the ones when the sound can be heard the oftenest.

Dangerous snowslips.

The dangerous avalanches come usually with heavy snowfalls. The snow is then heaped up to overflowing along slopes and in shallow basins, becomes packed and heavy with thawing and freezing, and finally splits across and loosens at the bottom. A slight cause may then start them—the wind, the report of a gun, even the bray of a donkey, according to the tale of Swiss guides. The detached part, perhaps, rushes down over some huge escarpment or cliff, where it is flung out in a white shower upon the air, and goes thundering down through the pines into the lower valley. These snow and ice avalanches are frequently seen on the cliffs of the Jungfrau falling in a veil or cloud of snow. Usually they are dangerous slides because they come in unexpected places and are as likely to bury a hut or village as a tree or a boulder. The shock of their fall, with its reverberation, is like distant thunder. Moreover, when the avalanche is very heavy with ice it is accompanied by much friction on the mountain side from which proceed violent whirlwinds of eddying snow. These winds are said to be violent enough at times to twist off trees at their roots and to be stifling enough to destroy life. But it would seem to require great weight and friction to produce such results.

The most dangerous and destructive of all the avalanches is more of a landslip than a

Ice avalanches of the Jungfrau.

Friction of the avalanche.

Landslips.

snowslip. By alternate freezing and thawing the ground beneath an ice-sheet is loosened, perhaps, down to the rock strata, and presently the whole mass breaks away, carrying down with it enormous boulders, trees, fallen timber—in fact, anything and everything movable. The swiftness and violence of this descent, the bounding and crashing of the huge boulders against the rocks of the slopes,* the wild leaps in air over cliffs and jutting ledges

Destructive nature of landslips.

are marvellous to behold. When the landslip finally spreads and collapses in the valley it possibly squeezes out of existence a beautiful lake or fills the whole valley with its débris. Such landslips are of frequent occurrence in the Alps, and the untenanted regions of the Rockies and Andes still bear abundant evidences, in the slashed pathways through the

Caucasian avalanches.

timber, of similar visitations. In the Caucasus and the Himalayas the avalanches are far larger and more violent than in the Alps, because of the greater accumulation of snow and ice; but they do less damage to humanity and are not frequently seen by the people of the Western world.

* John Muir has described how fire is struck from the contact and how vividly it shows in the night.

Though the glaciers and the avalanches cannot be said to work any direct good to humanity or the valleys into which they fall, the snows and ice that send them forth are by no means an unalloyed evil. They protect the mountain-top from the more formidable erosion by wind and rain, acting as a cloak and shield against the elements. To be sure, they are responsible for the ultimate pulling down of the mountain to the valley and plain, but they do not act with the swiftness of rain and wind on exposed peaks. The wearing away of the bare mountains below the snow-line is greater in proportion.

Nor can there be shown anything very beautiful about an avalanche, though if humanity knew just when one was about to come off there is no fine beauty of nature they would not desert to see it. It has the morbid drawing quality of all sudden catastrophes. The volcano, the earthquake, the cyclone, the avalanche are all of them catastrophes, accidents, attempts at readjustment of equilibrium or pressure. They represent nature not "red in tooth and claw" so much as nature helpless against the operation of her own laws. But, whatever the fury of the elements or

Readjust-
ments of
pressure.

however destructive the accidents, they seem of only local importance—mere surface ripples. The gates of brass and walls of jasper are set trembling for a moment but the great foundation is not shaken. The globe is made of sterner stuff than that. And nature's se-

Nature's
serenity.

renity is so supreme that it is scarcely ruffled, even for a moment, by tidal waves and cyclones, by the explosions of Krakatoas, by the thunders of avalanches in the Himalayas. Heat, light, and moisture still endure, and working together they soon repair any damage, heal over any break or abrasion of the surface. The livery of the earth is being continually worn and torn, but the great shuttle is forever weaving new garments and, through change, making permanence possible.

CHAPTER IX

THE SNOW-LINE

HIGH above the timber-line, above the barren uplands, comes the snow. Its beds and drifts and glittering caps are the first things to catch the eye. Down on the plains or the prairies, in the heat of summer, one looks up to see, perhaps many miles away, those patches of white against the blue, those cloud-like traceries between heaven and earth. Pale patches afar off, motionless, shadowless, colorless, almost formless, what phantoms of the sky they are! Yet these are the mighty pinnacles of crystalline rock forced upward from the bottom of the earth's rim, these are the eternal snows that endure from century to century.

The snow-line varies with the latitude of the earth's surface. In Greenland, near the pole, it is at the sea-level; in the Andes, under the equator, it is thirteen thousand feet or more in the air. It varies again with the seasons, which is, in its final result, the same thing as a variation of latitude. The Alps in winter have a snow-line at two thousand feet and in summer at about nine thousand. The Canadian Rockies vary but little from this, though the snow-line in the central Rockies is higher during the summer months.

Not sharply marked.

Just where the snow-line begins or ends is not so sharply indicated as in the case of the timber-line. On almost every high mountain slope there are gulches and ravines where the snow packs and lies the whole year round, though the neighboring slopes may be bare or covered with short grass and flowers. These patches of snow are usually the outliers of the great fields above them and sometimes are mere extensions down from them. They often gather on the northern sides of the mountains, and from melting and thawing or from the peculiar forms of their beds they take on odd appearances when seen at a dis-

Snow in rock fissures.

tance. Where snow has drifted and settled in a wide fissure of rock, which in turn has been broken by a cross-fissure also filled with snow, there may be the appearance of a white cross on the mountain side and the mountain itself be known as the Mountain of the Holy Cross. Again, on rough uplands or rock faces patches and clots of snow will sometimes appear in a broken line looking like an Arabic or Syriac inscription of white marble inlaid upon verd-antique or red porphyry. And, still again, one sees resemblances in these lodgements of the lower snows to human profiles, to witches, animals, ferns, or branches of trees —to almost anything fantastic or grotesque. *Fantastic appearances.*

The snow-drifts in the upper valleys are not clean-looking when seen close to view, nor are they very white seen from a distance. More or less débris from the hills and dust from the air have made them far from immaculate. Sometimes they are little more than a chocolate-colored lump, like the Nisqually Glacier on Mount Rainier, and from them may run a sand-colored stream. The snows on the southern peaks of the Coast Range are usually of this character, and through melting and freezing are more like heaps of rotten ice than fallen snow. *Gulch snow dirty-looking.*

But not so with the snow mantle of the far upper peaks in the great ranges. Nothing can exceed it in brilliancy. It is the whitest thing on the earth, save only the sunshine that makes it apparent to us. The minor poets have always used snow as a symbol of purity, and the sentimentalists, time out of mind, have *Purity of mountain snow.*

written about things "white as the driven snow," but it is to be feared that they usually have in mind the snow on the front lawn and the gate-post—the valley snow. Shakespeare, who is always so exactly right in his adjectives and comparisons, shows a wider and truer knowledge when he makes Ophelia sing:

Its white-ness.

"White is his shroud as the *mountain* snow."

The mountain snow is the highest and the last degree of whiteness. Nothing goes beyond it or can exceed it.

Yet Shakespeare's knowledge of snows was, perhaps, more a summary of universal opinion than any first-hand observation or perception. He alone said it, but many before him saw it and thought it. Perhaps that is why the name "white" is so often applied to the snow-clad ranges. Himalaya means "snowy," Lebanon is "to be white," Alp is from alb (white); Mont Blanc, the Sierra Blanca, the White Mountains all mean the same thing. The whiteness of the snows upon the peaks was, even from the beginning, so impressive that it stamped itself upon the name. Why not? What more expressive and truthful designation could be given them?

The word "white."

An explanation of the intense whiteness of the snow was offered long ago by Tyndall, and it has been generally accepted by the scientific world. It is, in brief, that the whiteness is caused by "the mixture of ice particles [in the snow] with small spaces of air." That suggests the cause; and the whiteness of sea-foam, of powdered sugar, of powdered granite, may be accounted for in an analogous way. But the explanation is not full enough, and, regarding the snow in particular, I shall venture a further statement of my own, set forth some years ago.*

Why the snow is white.

Snow is crystalline in character, and when examined under the microscope each separate flake is found to have prismatic edges that disperse the colors of the spectrum. These colors are the component parts of light—in fact, light disintegrated. The countless small prisms scatter the light into colors, but the mass of the colors taken together reunite into light. It was demonstrated in painting by the so-called impressionists, thirty or more years ago, that small stipplings of red, yellow, and blue placed upon canvas close together would throw out more light than a pure white

Crystalline character of flakes.

* *Nature for Its Own Sake*, p. 106.

ground or pure white paint. In other words, light recomposed from colors is stronger than light reflected from a neutral ground. The foam of the breaking wave is glittering white for a similar reason. It is filled with air bubbles and each bubble, again, is a prism of color. The multitude of colors taken in a mass recombine into white light like the prism edges of the snow.

The crystalline quality of the flake—the ice in it—makes possible the brilliancy and the dazzle of the snow, and yet the sharp contrast of the snow with its darker surroundings may have something to do with our appreciation of its whiteness. Alpine travellers, camped on the slopes of the upper peaks at night, have insisted, more than once, that there was a phosphorescence in the mountain snow, so luminous does it appear under the stars. It is brighter than the sky above it from which emanates whatever light may be. But how can a reflection be brighter than its source, a satellite more brilliant than its parent sun? Phosphorescence will hardly explain the luminosity of the snow. There is something wrong in the observation or the conclusion.

It is a fact that the snow is brighter than

the sky even at midday, but it is not a fact that it is brighter than the light that comes *through* the sky. We are misled by the way we look at it. Gazing upward at the blue, we inevitably look into the underneath or shadowed side of every air particle, whether dust or moisture or snowflake or mere intangible vapor; looking downward, we just as inevitably look into the high, reflected lights of every snow particle. The difference is simply between light cut off by shadow and light sharply reflected and enhanced by color combination; but that is practically the same difference that there is between the sun face of a mirror and the shadowed back of it. It is the sun face that we are seeing in the snow, the shadowed back that we are seeing in the sky. Hence the greater apparent luminosity of the snow.

Almost as beautiful as the light of the snow is the shadow cast upon it. Taken in large areas, nothing could be more delicate, elusive, and mysterious, without obscurity, than the pale shadow upon a snow peak like that of Mont Blanc or Mount Rainier or Popocatepetl. They are exquisite in their transparency and more lovely in their quality of color than any

Colored
shadows.

air blue of green valleys or rose and opal of desert peaks. For that the shadows upon snow have a hue is now no longer questioned. The Alpine climber who refuses to believe the vision of the impressionists, and who gets much shallow amusement out of ridiculing their discoveries and practises, can hardly lift his alpenstock out of the snow without noticing that the hole left by it has blue sides. It may be some time before he thinks of looking at the hole at all, but when he does he sees a hue that looks somewhat like his linen when it has been overblued in the laundry. That convinces him of the reality of the blue shadow on snow, and he then goes farther than he should, perhaps, by supposing that it is always and ever present. The new convert is usually overzealous.

But the blue shadow upon snow does not appear except under certain conditions. On cloudy days it does not appear at all—some of our young landscape-painters to the contrary notwithstanding. It requires, first of all, a clear sky overhead. This leads one to suspect that the reflection of the blue sky is thrown on the snow and is more apparent in the shadow than in the light because it is not bleached out by the sun's brilliancy. Very likely some of the blue in the shadow comes Sky re-
flection. from the sky because, even in the early morning, before the sun is over the horizon, some blue will show. But there is another, a contributing if not a principal cause for the blue shadow.

The most positive blue is seen only when there is a yellow sun in the heavens, and sometimes when there is a cold white sun the blue in the shadow is not apparent. It is a well-known scientific fact that colors cast their *Comple-mentary colors in shadow.* complementary or opposite hue in shadow, and the complementary hue of yellow is blue. Any one can test this at the lunch-table by using a white table-cloth instead of snow. With the ordinary light of noonday coming in at the windows there will be no blue shadows cast by glass and cup and bowl; but bring in a lighted candle, and the yellow flame will instantly color the shadows an exquisite blue.

This is not, however, absolutely conclusive *Effect of yellow light.* proof. If you leave the candle burning and close the shutters, shutting out the daylight, the blue shadow will as instantly disappear from the white cloth. The dinner-table at night may be ablaze with yellow candles, but

Shadows on snow of Mont Blanc.

the shadows of the candlesticks or water-glasses or flowers will be merely a colorless gray. The blue sky, or the peculiar light coming from it, certainly has something to do with the blue shadow, and yet, just as certainly, that in itself is insufficient. I have seen twelve hours of blue sky over Mont Blanc that produced not a trace of blue on the snow. The shadows of the snow-fields were dove-colored or a silver-gray—beautiful in their tenderness of hue as in their depth and transparency, but not blue. Again, I have seen several times, with a red sunset, shadows on snow that were greenish. The controlling factor seems to be the color of the sunlight. The combination of yellow sunlight with blue sky, or the light from it, seems to be necessary to the production of the pale-blue or lilac or purple shadow.

Alpenglow.

Reflection of the sky, however, may play some part, and that the snow is often a reflector we know from some more positive manifestations. The *abendglühen*, or alpenglow, that comes at sunset or after it, and is seen on all snow mountains when the sky is highly colored, is nothing more nor less than a reflection of the red or pink or purple in the twilight sky. This is direct reflection having nothing to do with the complementary color. The pink or purple or golden hue on the snow appears in the light and not in the shadow. It is similar to the reflection from clear ice, only it is not so sharp in definition.

The bare rock of mountain peaks will often reflect an alpenglow quite as effectively and sometimes more delicately than snow. The bald top of Grayback (about 11,600 feet), the highest mountain in southern California, seen from Dry Lake on the desert side, shows in the summer sunrise first a pale gray, then a delicate silver followed by salmon-pink, then a pale yellow changing into gold. When the sun rises higher the rocks become silver-gray and the shadows in the neighboring gulches and canyons become a gun-metal blue. This effect is practically repeated at sunset, but with less delicacy and variety of hue—that is, seen again from the desert side. Other peaks of the desert are more vivid in their sunset effects, as I have already suggested, and the peaks in any range occasionally take on unexpected colorings.

All these colors and shadows exhaust one's adjectives and still get away undescribed.

Delicacy of their color.

Their delicate beauty is more like the tints of the eastern sky at sunset. Indeed, so subtle and intangible are many of them that they are seldom observed by tourists and mountain-climbers.* The bright pink or gold of an alpenglow seen looking up to some peak from a darkened valley is, of course, remarked because it is so pronounced, but the blue and dove-colored shadows, though often looked at, are seldom seen. Just so with the cold

Blues and purples.

blues and purples of the snow crests in the early morning before sunrise, when the mountains are shadowless and the peaks swim against the sky like northern icebergs against the aurora. After sunrise, if the sun or sky is of a yellow tone, the snow beds will often reflect it in a pale saffron somewhat after the manner of the evening alpenglow. The snow is, indeed, as capable of direct reflected

---

* I have laboriously gone through many volumes of Alpine Club publications, with tales of ascents and descents, in the hope that perhaps some of the climbers would talk about something besides their hair-breadth escapes and physical worries, but I have had small success in my search. The average ascent-maker seems to have better legs than eyes. He sees little save the man ahead of him and the peak above him. He is doing a stunt—not seeing a vision. People like John Muir and Leslie Stephen are rare in mountain literature.

color in the high light as of complementary color in the shadow.   They may both appear at one and the same time.

> "Hues in the light and hues in shadow fading,
>   Splendor from sunset glows,
>   Fields of blue into mauve and lilac shading,
>   Tintings of gold and rose.

> "Was ever known such light by hue attended,
>   Such wondrous radiant light!
>   Was ever line with line so subtly blended
>   As in this patterned white!"

Moonlight among the snow peaks is a marvellous nocturne in blue and silver.   Everything has the soft gleam of the wan white moon.   Even the stars are silvery white and glow from their deep-sunk beds with a preternatural splendor.   Against the blue, against the stars, the snow peaks appear sharp-edged, clear-cut, apparently quite flat.   At times they seem to shift slightly, especially if the moon is in the line of sight; but we know their steadfastness, their absolute repose.   "I beheld the mountains, and lo, they trembled, and all the hills moved lightly."   It was a vision rather than reality that the prophet beheld.   Yet the shaking atmosphere is at times responsible for a "tremble," and again, at times, the clouds "move lightly" and give the appearance of moving peaks. But the peaks sway as the globe sways, not otherwise. To ordinary eyes they stand with their white summits thrown like high lights on the blue, the embodiment of stability, permanence, and might. No other feature of the globe has their loftiness and their

aloofness. Silent they rest in the thin upper air, above the noise of the world, above the dust of conflict, almost above the very elements themselves.

The forms of snow are quite as wonderful as their colors or shadows. And the snow is always forming or being formed. Great troops of clouds trail across the upper sky and let down sheets and whirls of white flakes upon peak and valley. Far into the timber-

line, even in summer, the soft crystals fall, until the pines and the bare uplands above them look as though they had been sprinkled with powdered sugar. Higher up, where there is less melting, the snow falls thicker and faster, forming in great, cloud-like banks or catching and resting fluffily upon vast slopes and mountain-backs. When the snowfall ceases and the sky clears, perhaps the wind is still up, and from the valley the mountains seem to be smoking or steaming, so high and wild are blown the clouds of snow. They are tossed about the peaks, swirled around the stony needles, huddled in patches along the broken ledges, banked in drifts across the ravines and crevasses. The fury of the wind around the upper peaks makes wild confusion among the flying snows.

Banks and drifts.

But out of it comes incomparable beauty. For when the snow is blown about in such furious fashion the flakes lose their sharp edges, they grow smaller and rounder, and begin to pack closer and harder as they drift. They become

Bolted snow.

> "The fanned snow
> That's bolted by the northern blast twice o'er."

As a result we have drifts and banks somewhat like those of wind-blown sand in the desert, only more delicate and perhaps more fantastic. And of huge size and sweep. The snowfall on St. Gothard is between forty and fifty feet each winter. From that one can imagine that the drifts must be as deep as a small valley and as wide as the shoulders of

Snowfall on St. Gothard.

the peaks are apart. At times crevasses, ravines, and whole valleys are obliterated by drifts, and the only way one recognizes a valley is, perhaps, by a graceful downward droop of the snow between peak and peak.

These rolls of the deep drifts are supremely beautiful as line and form, marvellous because apparently designless patterns of nature; and yet just as perfect after their kind as the starred snowflake and the colored crystal. Ruskin, to whom is due perhaps even greater credit for what he has pointed out in nature than in art, has written rapturously and delightfully about them.

"In the range of inorganic nature I doubt if any object can be found more perfectly beautiful than a fresh, deep snow-drift seen under warm light. Its curves are of inconceivable perfection and changefulness, its surface and transparency alike exquisite, its light and shade of inexhaustible variety and inimitable finish, the shadows sharp, pale, and of heavenly color, the reflected lights intense and multitudinous, and mingled with the sweet occurrences of transmitted light."

This is not only enthusiasm but truthful observation; and it might be noted in passing that so far back as 1843, before ever the impressionist and his blue shadow were heard of, Ruskin had observed that the shadows on snow were of "heavenly color"—that is, blue. Only a few years later, in Massachusetts, the not too observant Lowell noticed the blue shadow also, and wrote about it.* The impressionist did not invent it; he did not see it until others had called attention to it. The old masters of Flanders, such as Thierry Bouts, continually painted blue shadows in the folds of white robes. The knowledge (and occasionally the practise) was not unknown to the Renaissance Italians. *Ruskin on blue shadows.* *Old masters knew the blue shadow.*

After a heavy snowfall in the mountains it not infrequently happens that the temperature rises and a south wind or (in the upper Rockies) a chinook blows for several days. The snow melts on the surface, becomes hard by freezing at night, and gathers to itself a shine and glisten under sunlight and moonlight. The festoons that hang down from the sharp peaks and the cornices of snow that extend out over the edges of the precipices are now congealed in graceful forms, and perhaps the cornices are fringed by drooping *Melting and freezing of snow crust.*

* See his essay, *A Good Word for Winter.*

icicles that have a beryl-green or sapphire-blue color in the sunlight. The upright walls of the peaks may again be hung with crystal pendants and the whole mountain range glitter like spun glass. The condition of the snow is now favorable to the most gorgeous effects of the alpenglow, and this is the moment when the snow splendor is fairly dazzling. Not Mont Blanc alone but every high mountain becomes, at such times, a monarch

> "With a diadem of snow."

The crown of glistening crystal is, indeed, the supreme touch of heavenly light upon "the snow-shining mountains."

And yet in severe weather, in very cold climates—in the high points of the eastern Rockies, for instance—there is still a further enhancement of this mountain beauty by an air that is sometimes filled with hoarfrost or ice-dust. Every particle of moisture is frozen out of the air by the intense cold—is turned into tiny ice spicules that float in the air and glitter like so much diamond-dust. Moreover, the crystals in the air are occasionally illumined in the early morning not from one sun but three! For in cold climates, with snow on the ground, the sun often comes over the eastern horizon flanked by sun-dogs (parhelia), one on either side, and each one almost as brilliant in light as the central orb. Then there is the glisten of a fairy world—the dazzle of a world all splendor. The glitter from the million patterns of ice tracery and snow crystals is bewildering, almost blinding.

Sun-dogs.

Up through this drifting crystalline atmosphere lift the white pinnacles and minarets of the peaks. Standing along the edge of the great Blue Unknown, they seem like watch-towers between heaven and earth. What a wonder-pattern of white upon blue! What do they mean? Nothing that we can know with any certainty. What do they look? The most splendid vision that human eyes have ever seen.

The lifting pinnacles.

CHAPTER X

SPINES AND WEDGES

THE snow would seem to be the last word about mountain height. But there are still needle-pointed rocks beyond it, splintered peaks that thrust through the white cloak and lift into the sky.

> "Where the arching Bowl of Blue
> Into outer space breaks through,
> Where the Milky Way is strewn
> And the night is starry noon,
> There the splintered peaks aspire,
> Lifting higher, ever higher,
> Into ragged shaft and wedge
> Far beyond the round world's edge."

The highest needles and the sharpest crests are not snow-tipped. Such rocky points are too perpendicular to allow lodgement for banks and drifts, and they usually remain bare as an obelisk or the apex of a pyramid. They are, indeed, nature's tall shafts set up as landmarks in the wilderness, and they tell a history just as truly, and far more weighty, than,

180

say, the rock-hewn monuments along the Nile. To understand them we shall have to look up their origin in geology.

In the original laying down of the rock beds the normal inclination must have been horizontal or possibly sloping; and the beds must have originated and been laid down by and through certain elements, such as fire, water, and wind, and solidified by compression. *Inclination of rock beds.*

(*a*) The igneous or fire rocks were made by lavas intruded through seams, or by overflows of lava from volcanoes, or possibly by proximity to certain underlying molten masses of the earth's interior. The fusion of the different layers by these fire streams produced further lavas, tufas, glassy stones like obsidian, porphyries, granites, greenstones, basalts, and trap rocks in general. Molten rock is different from molten metal in that the latter returns to itself after cooling, but the fusion of rocks produces a solution of one silicate in another. Each compound crystallizes out as it cools until the whole is solidified. Liquid rock is like glass or furnace slag and crystallizes in substantially the same way. Generally speaking, then, we may say that the igneous rocks are unstratified crystalline rocks. *Igneous or fire rocks.* *Crystalline results.*

Aqueous rocks.

(*b*) The aqueous or water rocks are of secondary origin—that is to say, they are formed from the grit and fry of the original mantle rock. These minute wearings are finally carried by wind and water down to the sea or to the lakes, the plains, or the deep valleys where they form in beds. Lying in these beds under great compression, they finally become cemented in compact strata and make up shales, sandstones, pudding-stones, and the like. Lime, magnesia, soda, potash leached out of the surface rocks are also carried by the rivers down to the sea, where the lime is extracted to make the bones and shells of sea life. The bones are finally deposited on the sea floor in great quantities and there help to make up a lime mud which eventually becomes limestone. Generally speaking, then, it may be said that the aqueous or water rocks are not crystalline but sedimentary in structure and stratified in form.

Stratified results.

The soft rocks.

(*c*) There are soft rocks formed also by winds (dunes and loesses), by glaciers (moraines and till), and by organic deposits (coal, peat, lignite); but these do not figure in the creation of mountain peaks and may be dismissed from present consideration. There is, however, still a third kind of rock to be spoken of, produced by what is called metamorphism—that is, by a refusion and remaking of igneous and aqueous rocks that have been crushed, broken, or disintegrated. The muds that become shales and limestones, the gravels that become conglomerates, the volcanic blocks and coarse fragments that become agglomerates are metamorphic rocks in one sense but not in the sense we are for the moment to consider. *Metamorphism.*

When sedimentary and igneous rocks are heaved up by a folding of the earth's crust there is tremendous thrust and friction, and these, in conjunction with moisture, result in high temperature. The heat is so great that it seems at times as though actual melting of the rock were possible or even probable. With the heat a change or transformation in the character of the rock takes place. From sedimentary or conglomerate form it turns into schistose or crystalline form. The sandstones and granites pass into quartzites and gneisses, the shales become mica schists, the limestones turn into marbles. Sometimes the transformation is so complete that the original character of the rock cannot be made out. This is the process called in geology regional *Friction and heat.*

Transformation of rocks.

metamorphism, and the result of it is the schistose or crystalline rocks that appear in the needles and crests of the highest mountain peaks. How or why they appear there is something that requires a further paragraph of explanation.

When the earth's surface is contracted by the cooling of the globe a lateral thrust results which squeezes the surface into wrinkles or folds. When the thrust is slight the surface merely bends up in a quarter or half circle arch. This is called an *anticline,* and the crust-waves may appear somewhat like the long ground-swells of the Pacific Ocean, only much larger. When the curve is downward instead of upward it is called a *syncline,* and the resemblance is nearer, perhaps, to the hollow of the wave than the arched top. If the thrust is increased the folding is closer and the arches are narrower. There is a bend both up and down, with high, sharply buckled crests and deep, narrow hollows. This is called an *isocline.* Increase the lateral thrust still more and the folding becomes narrowed at the bottom and flared out at the top like a fan. This is called an *anticlinorium.* It reveals the most violently bent and broken folding of all, and results not only in a snapping of the arches at the top but oftentimes in overfolding or the thrust of one fold over another in a hopeless confusion of splintered edges. Aside from these there are many complex forms of foldings to be found in every mountain range, but the ones described are sufficient for our understanding of the different kinds of mountain rocks, how they were formed, and how they were placed in the mountains—some in hidden beds, some in huge supporting buttresses, and some in exposed peaks.

Complex foldings of crust.

If, by way of illustration, we now examine in profile the foot-hills of a mountain range, or the lesser ranges themselves, we shall find, generally speaking, that the lateral thrust of the earth's folding has, perhaps, not been great, that the bending has been possibly of an anticlinal or synclinal nature, that there has been no great heat, and, consequently, no marked metamorphism or change in the rock structure. The beds bent upward or downward are of sedimental or aqueous origin and are usually shales, sandstones, limestones, or conglomerates, the breaks in the strata are not violent, and there are no needles or crests of importance.

Foot-hills as illustration.

Outlying mountains.

The outlying mountains that range themselves along or about the high central peaks we shall find fairly good examples of the isoclinal bending of the crust. The thrust has been strong enough, perhaps, to produce in spots metamorphic rocks, to change some of the sedimentary beds into gneisses and schists, while leaving other portions, like the limestones, unchanged.

Ragged edges of beds.

The folding has been severe enough to snap some of the arches at the top and leave the ragged edges of beds and layers exposed to view. These edges frequently appear on the summits of the secondary mountains in sharp ridges, in fantastic parapets or towers or domes of rock; but owing to their mixed nature—their half-fired quality—they are not hard enough to endure sharp-edged for any length of time.

Wearing of the edges.

The forces of erosion wear them smooth or honeycomb and crumble them, or they are covered by lichens so that the dome soon sinks into a knob and the crest becomes merely a smooth shoulder. This is, however, a general statement subject to exceptions, for, as every one knows, some of the now lower mountains are still bristling with sharp schistose or crystalline ridges.

The very highest peaks are made up of the

hardest rocks of all. The violence of the fan-fold not only generates great heat and changes the rocks of whatever name and nature into crystalline forms, but it forces the arches high up in the air—high enough to make the topmost peaks of the Rockies, the Alps, or the Himalayas. Further, the enormous thrust of the folding results not only in the stratified beds being bent to the perpendicular and raised on end like planks placed upright against a wall, but in the tops of the arches being broken through and the ragged edges of the beds pushed up in spines and needles. This is the climax in mountain-building and means the hardest rocks on the highest peaks splintered into fantastic minarets by the wildest waving of the earth's strata.

The needle-points, such as the Aiguille du Dru, the Aiguille des Petits Charmoz near Chamonix, the pinnacle of the Weisshorn, the shaft of the Matterhorn, are accounted for by the upright position into which the strata have been lifted. The gable-shaped peaks and the ragged crest of the Schreckhorn or the ridge of the Wetterhorn are produced by portions of the strata having been broken out by the winds and rains so that what re-

mains appears often as a pyramid or a wedge
or a ridge.* The most amazing illustration
of this is the shaft of the Matterhorn, though
just how or why it happened to be left in
its present fantastic form has not as yet been
satisfactorily explained.

Of course the outlined view of a mountain
range against the sky will not always disclose
the pyramid or the wedge or the needle. At
Belluno, looking west at sunset, the mountain
peaks look like thunder-heads in the sky or

mounting sea waves. The Dolomite peaks of
the Schlern and Rosengarten, looking east
from Botzen, or the Organ Mountains in New
Mexico have needles thrust up that resemble
the pipes of an old-fashioned organ; and the
Canadian Rockies make up a wild expanse of
snapped-off bends and bows with brittle edges
—countless peaks with sloping backs on the
west and sharp, broken precipices on the east.

* The Alps present the most violent folding and
huddling in the crust materials of any of the famous
mountain ridges. In the crushing process not only
metamorphic rocks have been created but the older
igneous rocks of the crust lying under the sedimental
deposits have been forced up to the surface, pro-
truded, splintered into needles, and in many cases
torn off and pushed for many miles over and above
the tops of the younger sedimentary rocks.

The main sierra running through Mexico, or the Carpathians seen from Sinaya, have a similar variety of form. But under all the variety runs the principle of the needle or the wedge or the ridge—the rock bed tilted up on end or at a sharp angle, with its once arched top shattered and its rugged, crystalline layers forced outward or upward in the air.

Mexican sierras.

Many of these splintered spires and needles have survived and still exist because of the hard, gneissic quality of their rock. Crystals cannot be disintegrated like sandstones nor crushed so easily as conglomerates. Nevertheless, they are, in the long account of time, worn down in spite of their adamantine character. Heat and cold expand and contract them; wind and rain wash and wear them. At a distance, looking up at the peaks from the valley, the edges appear as keen as a knifeblade and as pointed as a pike; but when you climb up to them and see them close to view they are almost always rounded. They are all of them "weathered," and some of them are entirely covered with gray and black lichens.

Hard, gneissic needles.

Weathered peaks.

But even the lichens do not hide the youthfulness of these peaks. They have survived

Youthful
ranges.

through their crystalline hardness; and yet
it must not be forgotten that they have not
been exposed for long. It was only geological
yesterday that the Himalayas, the Alps, the
Andes, and the Rockies were born. Indeed,
the geologists think that they have not yet
run their course, not yet come to maturity,
and that they are still rising, folding, and
forming. Their birth-time was not earlier
than the Tertiary Period. The inference from
that is plainly that the sharpest needles and
the most angular crests are the youngest
and the latest of the broken foldings and the
splintered strata.

Local color
of the
peaks.

So young are they that they have not yet
had time to gather much local color. The
lichens lend a tone but not a hue, sometimes
the oxidizing of iron or copper gives them an
orange stain; but as a rule the local coloring
of the high mountain peak is merely a dark
gray or a purplish brown with occasionally
a tinge of silver or saffron. Sometimes, again,
there are tops and towers of a chalky hue that
throw off beautiful pale tints of heliotrope and
lilac; but, once more, the local color of the
peaks is a negligible quantity in that high field
where even the blue sky sometimes holds by

its value as illumination rather than by its splendid hue.

This is not, however, true of the peaks when they are reflecting color and light from the sky or air. The bare porphyry shafts of the Colorado Desert are almost unbelievable in their fire reflections. At sunset the peak of Baboquivari, along the Arizona border, is often a hot-iron red, and the needles of the Mexican mountains in summer assume strange gamuts of reds, pinks, golds, and lilacs. I have never seen anything so intense in the Rockies, the Alps, or the Caucasus. In the Dolomites, seen from San Martino, the spines and needles are at times golden or pale yellow or gray-silver under sunlight, but the color is far from brilliant. When powdered with newly fallen snow there is produced an effect of pale orange and silver on white, and in the morning this effect may be enhanced by wandering patches of mist which, shot through by the rising sun, are like golden fleece. The effect of snow swirls and drifting snow around the needles under morning light is very similar and even more wonderful in luminosity and glittering splendor.

The best "view" of the high peaks is always obtained from the valley looking up. It is

Reflected color of bare rock.

Dolomite spines.

Snow swirls about spines.

usually an imposing view and often a spectacular one. The lift and the bulk of the peaks give them sublimity; their repose and serenity lend them dignity and majesty. They have not the graceful contour of the lower hills and do not please the eye by one line flowing into another in uninterrupted sequence. In fact, the saw-toothed effect of a lofty range seen in profile may be anything but agreeable

to the sensitive retina. But all that is swept away by overwhelming mass and grandeur. We are face to face with the crumpled crust of the earth—no more than a mere wreck of matter—but it is upon such an enormous scale that even in its ruin and confusion it is wonderful to behold. The Canyon of the Colorado is just as much of a wreck—a great scratch in the crust—but again we have sublimity as a result of its tremendous expanse.

The view from the top of the mountains— the look across many peaks in panoramic arrangement—is the one the Alpine tourists usually extol as "magnificent" and "unsurpassed." So it is. You are on the roof of the world, with snowy gables, chimneys, and spires all about you. It is usually a tumbled and tossed landscape, "ringed and roofed in azure," with huge fields of snow and shoulders of ice, and looks more like a frozen polar sea covered with icebergs than any portion of the earth. Such a scene meets the eye from the top of Monte Rosa, the Gorner Grat, or even the Righi. *From Monte Rosa.* The look over the tops of the Canadian Rockies from Mount Balfour shows perhaps less abruptness in single heights or depths but is not the less a wild and waved landscape. The whole region is an uplifted pene-plain of sedimentary rock fifty thousand feet thick or more. It was originally folded *The Canadian Rockies.* and faulted by a squeeze from the west which was resisted by the underlying strata of the flat plains to the east. In its upward bend it broke into troughs with open crevices which widened as they became the beds of drainage streams. The upheaval seems to have been evenly felt, judging from the rather uniform height of the remaining peaks. The view from Rogers Peak in the Selkirks is about the same as regards the uniform height of the peaks.

All these views seen from high peaks when the snow is lying thick and heavy bear *Chaos of snow-fields.* some resemblance to the south-pole ice barrier as photographed by Lieutenant Shackleton. They are a little disturbing in form because of their chaos; they are bewildering in their light because of their snow mantles; and, as for their color, no one looks for it or expects to find it. Intense light and disrupted form are too prominent for anything like color to attract attention.

The view looking down.

The view looking down from the peaks is again one that is popular with the casual climber, perhaps because it is more or less of a distortion. It is somewhat like seeing the earth from a balloon. Light, shade, color, perspective are all reversed if not destroyed. Topographically, it may be interesting from a great height to trace the twist and bend of the valleys and to watch the glittering thread of a river winding downward to the sea, as, geologically, it may prove instructive to note the different river benches and the part played by erosion in their making; but, picturesquely, the looking downward is something of a freak.

The distorted impression.

It amuses for a time but does not satisfy, because it quarrels with all our usual experience in landscape. Perhaps, if the climber would look up instead of down, he might see something of unusual beauty in a shade of blue bordering upon purple; but unfortunately he seldom looks that way.

One wonders at times why the alpinist undergoes fatigue and cold, besides risk to life and limb, in climbing among the lofty peaks. He talks of glorious views, fine air, and a fine physical feeling; but at heart, perhaps, he likes better the name and fame of having "done" such and such peaks and having stood on dizzy heights where only the Whympers have set foot. The bare needles of the lofty ranges are not spots that any one may reach in a day's journey. They are accessible only to the very few, and the alpinist is one of the elect. It took a Hannibal, a Cæsar, a Napoleon, to cross the lower Alpine passes, and the great Alexander was turned back from India by lesser difficulties. The conquerors who rise to mountain heights must be the strong of heart.

But, whatever the difficulties, in its last analysis the present-day climbing of perpendicular walls of rock, the zigzagging in chimneys, the drag by ropes, the crawl by ledges and niches, the creeping along knife-blade edges are merely Alpine Club stunts in an old game of "follow my leader." It is the dare of it that makes up the game, and the beauty of the peaks and skies forms small part of it.

On the contrary, there have been many soli-

The Alpine climber.

Difficulties of high ascents.

The dare of climbing.

tary mountain-climbers who have never talked their ambitions and never published their achievements in club-books—people who, perhaps, are not overfond of the view, care little about the physical exercise, and care nothing at all about the game or the fame of it. Why should such people climb? Can it be that they are drawn to the lofty heights by their wildness, their aloofness, their loneliness? No one owns the pinnacled peaks; no one lives

up there. The mountain-tops, whether spired or domed, are out of civilization, out of the world. Is there attraction in this remoteness —a fascination in this outward look into the wilderness of space?

The true mountaineer is like the sailor. Neither of them can say why it is he is

drawn to the great open spaces. The longing for the sea comes to the man in the street and he leaves family, friends, and fortune, to take ship and toss about in stormy waters. He cannot tell you why he loves the element, but the lure of it is more to him than home. So it is with the mountaineer. At times a restlessness and a longing for the mountains come to him and he must climb—the higher the better—and be alone with the elements. Perhaps it is the same longing for altitude and aloofness that comes to the eagles when at times they soar far up against the blue, wheeling for hours aimlessly in space.

This is something different in impulse from the misanthropic and the temporarily world-weary who like the hills or the mountains because they can get away from the mob. Bismarck used to declare that he was never so happy as in the mountains alone with his dogs; which, perhaps, meant merely that the human problem was for the time abandoned. The misanthrope, and even at times the healthy-minded, may, however, take another point of view and see matters in a more serious light. For to be high up on the snowy summit of the world is to turn existence back to the first days of creation. You stand with earth below you, sky above you, light around you. The primitive elements alone command you. All else vanishes. Time goes out and space comes in. The life down below becomes trivial and inconsequential. Humanity, if thought about at all, seems a blemish, and civilization a blunder. The wealth, the fame, the art you have so long striven for, and perhaps attained, how useless they appear! The knowledge so crisscrossed with error, the mentality of man so worthless in a crisis, how empty they seem amid these mysteries and vastnesses! Thinking is reduced to a mere vague wonder and emotional feeling is all that remains to us.*

How often that feeling in the presence of the great elements has expressed itself in a mist of tears and a choking in the throat! The high-blown pride of the human breaks under him just here. His reason deserts him and the religion of the Garden comes back to him. There in the high mountains, which were God's first temples more truly than the groves, he forgets to pray for himself, but has rapturous praise for the Power that planned and the Hand that wrought. He is back to a primitive faith from which he never should have wandered.

* Byron declared that to him the mountains were "a feeling," while Addison thought them merely "irregular misshapen scenes." What strangely opposed view-points! Addison is almost incomprehensible, but how readily one understands Byron! Hear him, in *Don Juan*, declaring:

"My altars are the mountain and the ocean,
   Earth, air, stars—all that springs from the great
      Whole
Who hath produced and will receive the soul."

CHAPTER XI

BLUE AND SILVER

THE blue of heaven! What a mighty depth it is! You rise to it as you creep up the mountain side, but it keeps receding from you, growing deeper instead of lighter. If you could rise high enough, say fifty miles, it would finally disappear in purple darkness—in a blackness of space set with blue stars that never twinkle and a violet-blue sun the rays of which would cut like search-lights through the dark. The upper space must be a void of darkness, for there is no tangible atmosphere there to break the light and make it apparent. Science has long been quite sure that the blue sky does not extend upward unendingly but, on the contrary, is only a circle of air or a translucent envelope that surrounds the globe and diffuses the sunlight. Fifty or a hundred miles is supposed to be the limit of its height or depth.

Darkness of space.

Limit of the sky.

This is something of a guess, to be sure, and yet it admits of some practical demonstration in the mountains themselves. Seen from the valley, the sky looks pale blue and the sun may have a tinge of topaz about it, but when we rise ten thousand feet on the mountain side the sun begins to look white and the sky has darkened in its blue. At fifteen thousand feet—at the top of Mont Blanc—the sky is the color of a dark-blue gentian with a purple note in it, while the sun has the violet tinge of

an electric arc lamp. From the top of Mount McKinley (20,000 feet) the Rev. Dr. Hudson Stuck noted that the sky "took a blue so deep that none of us had ever gazed upon a midday sky like it before. It was a deep, rich, lustrous, transparent blue, as dark as a Prussian blue, but intensely blue; a hue so strange, so increasingly impressive, that to one at least it 'seemed like special news of God.' . . . We first noticed the darkening tint of the upper sky in the Grand Basin, and it deepened as we rose."

At such a height as Mount McKinley the air becomes thin and the diffusion of the sun's rays grows less in so marked a degree that the sky fits up close around the disk. The stars, too, that from the valley go out with the morning light, are now sometimes visible at noonday. The air is not dense enough to hide them. As for the sun-shafts, they are so straight and strong that the climber suffers from sunburn on face and hands, and needs his smoked glasses to temper the dazzling reflection from the snow.

Even from lesser mountain heights, from, say, Mount Whitney or Mount Rainier, what an inverted bowl of blue is suspended above us in space! No precious porcelain out of China or Japan ever had such quality of color. It is not a tone under a glaze, but a vast depth—a sea of ether in which the eye wanders and loses itself in a mystery of infinite hue. As you look into it how empty seem all the formulas of color harmony devised by art and artists! That famous palette of Titian's which has so racked the modern brain with its composition, how petty it seems beside this vast blue monotone of the upper sky!—almost as petty as the Greek line of the human form compared with the rugged outline of the Matterhorn or the great swing of the horizon circle.

This cold blue color-scheme of the high mountain sky does not resemble anything seen in the arctic regions, except superficially. At the poles there is an atmosphere laden with moisture which at times produces vivid reds and yellows, but over Monte Rosa the thinness of the air usually limits color to the blues and violets at the far end of the spectrum. In the morning, when the sun comes up, there is usually little warm coloring resulting from it. The tip of the Matterhorn, perhaps, reflects a blue-silver, the Lyskamm and the Breithorn answer a few minutes later in tones of violet, the snow-fields put on a violet glow; but there is no riot of colors. Even at sunset the cold hues predominate, though occasionally the sky warms up with golden fire along the western horizon and the blue is green-tinged.

Our point of view usually determines the appearance of the sky. The twilight that looks so gorgeous in hue from the plains or the valley may appear less gorgeous when seen from the high peak. The warm red or yellow notes are made possible by the thick layer of atmosphere lying low down along the earth through which the sun's rays pass and are dispersed into color. The layers of air above the mountain-tops are very thin and disperse color meagrely, though occasionally one sees orange at dawn and fire-reds at sunset.

This thinness of the upper air is marked

when one is camped at night, perhaps, in an amphitheatre of snow near the peak. The sky may be clear and the moon at the full, but you may have to look about the sky to find the latter. Seen from the high mountains, it seems small, and though very bright it has little diffusion, and perhaps less intensity than the stars. Of course, it is not orange-hued nor yellow, as we see it from the valley, but very white, without a halo, and without rainbow rings. It is never spectacular, never romantic, never even commanding in presence or light seen from the high mountains. As for the stars, they are set in constellation patterns of great brightness thrown upon a blue-black ground. Rising toward them the violet-tinged peaks lift upward and upward and the world below is lost in a vast depth of mystery and gloom. You are camped half-way between heaven and earth and are, perhaps, awed by the unreal character of the light or overcome by the immensity of the solitude.

A strange feeling comes to one up there with the peaks and the planets—a feeling and a realization, perhaps, of

"The silence that is in the starry sky."

The moon.

The stars.

Poise of the universe.

It comes with the thought of the great poise and restraint of the universe notwithstanding its ceaseless activities. The stars in their courses, the planets in their orbits, the suns with their spinning satellites, how silently, easily, and yet inevitably they swing! What a mighty system set in motion apparently with no effort and kept in motion with no

The still-ness.

friction and no sound! The stillness is so intense that it may be heard in a hum of distance—a hum that comes out of the unfathomable silence of space. With such an immensity spread out before one, what wonder that the imagination breaks down and feeling becomes tense—even uncomfortable!

Peaks as cloud-makers.

Coldness of light and color is the usual feature of the upper peaks under a clear sky; but the mountains are great condensers of moisture-laden air, great cloud-makers, great rain-makers. Much of the time the snow peaks are shrouded in warm gray-and-silver clouds, the blue sky is shut out, and frequently the whole range will be lost to view in showers

The rain sheet.

of falling rain or waving sheets of snow. A local storm often takes place below the snow-line while the peaks above the clouds are still glittering under the sun or the stars; but a wide-spread rain of several days' duration brings with it a mantle of silver-gray that envelops the whole range. This mantle is often very beautiful in color, but is usually unseen or at least disregarded by almost all manner of men.

A night sky over a high mountain range is usually quite free from clouds. With the setting sun the troops of cumulus and nimbus sink down, disintegrate, disappear; but by nine or ten of the morning, when the sun warms the air in the valleys and the currents have begun to rise along the mountain side, the clouds reappear and begin forming, sometimes along the slopes and sometimes out in the open spaces. At the start the cloud is very diaphanous, but it soon gathers and grows on its top and sides until it becomes so large that it, perhaps, breaks in two, and then each half continues to grow on its own account. After a time the top of each cloud has a heaped-up look, the bottom of it a flat, sawed-off appearance, and the sides of it are marked by masses of light and shade. The cloud grows on its top and loses, by evaporation, at its base. The flat look of the base is the line of evaporation.*

Clear skies at night.

Rising and forming clouds.

* For further explanation of why clouds float and how they are formed, see *Nature for Its Own Sake,* Chapter IV.

Clouds in Alaska.

The forming and coloring of these clouds in the morning sunlight are seen to great advantage in a moist climate like that of the Alaska coast, where the mountains come down to the sea and the ranges are surrounded by sea air. I have never seen such superb masses of clouds, sun-shot and blue-shadowed, drifting in the valleys of the high mountains, as along the coast of Alaska. In the morning there are tones of heliotrope, lilac, and violet, making a cool opalescence; in the evening the whole mountain range as well as the clouds are splashed with yellow, gold, scarlet, and tempered by purples into a warm iridescence. Nothing in the tropics or the Mexican mountains can surpass the splendid color effects of

Cumulus.

these clouds. They are only ordinary cumulus clouds, but they declare the glory of God just as truly as the firmament showeth his handiwork.

As for the appearance and the unusual forms that clouds sometimes take in the mountains, they are easily explained. In the deep mountain valleys the moist air becomes heated in the early morning and begins rising. With every three hundred feet of rise the temperature sinks one degree, which means that at fifteen thousand feet the temperature would be fifty degrees lower than at one foot. At a temperature of 60° Fahrenheit, which is about that of the average mountain valley, each cubic foot of air is able to hold in invisible form 5.87 grains of vapor. As this cubic foot of air rises it grows colder and its ability to hold invisible vapor grows less. When its temperature is reduced to 32° Fahrenheit, and it is perhaps ten thousand feet up, it is able to hold only 2.37 grains of invisible vapor. The difference, then, between 5.87 grains and 2.37 grains is crowded out, condensed out, of each cubic foot in the visible form of mist or cloud. *Condensation.*

This active principle of condensation explains readily enough how and why warm currents of air creeping up the mountain sides, or rising straight up from the valleys, change into clouds when they reach the cold upper regions. It also explains the unusual appearance of certain clouds that form and hang about or above mountain peaks—so-called "bonnet" clouds. The cone-shaped top of Popocatepetl almost always has a ring about it or cap of cloud above it. The winds are always breaking it and blowing it *Bonnet clouds.*

away on the sides and top, but the cloud is always being renewed by the ascending currents of warm, moist air so that it looks stationary and intact. At night, when the ascending currents of air fail, the cloud sinks down and disappears. This type of cloud is seen in connection with many high peaks.

Banner clouds.

Somewhat similar to the "bonnet" cloud is the so-called "banner" cloud. It is formed by similar ascending air currents which perhaps rise along the warm, sunny side of a mountain and meet with no cold air until they pass out at the top from the protection of the peak. They are then struck by the cold wind, are condensed instantly into cloud, and are almost as instantly blown away by the winds, save in the lee of the peak, where there is pro-

Odd shapes.

tection. They gather under this lee in the shape of a burgee, or pointed flag, and, seen from below, seem to wave out from the peak like a streamer or banner. This cloud is often attached to sharp-pointed peaks and seems with some of them a permanent affair. But it is, like other clouds, continually forming and breaking and having its edges cut by the winds. The banner portion of it is merely the protected residue.

The warmer the weather, and the higher up *Rising warm air.* the region of cold, the higher will the valley air rise before it condenses. As a result, in summer weather or upon hot days at any time, condensation will take place only when the air currents are, perhaps, many thousands of feet above the valleys—thousands of feet above the peaks. The cloud form then produced is the towering cumulus, or what we call the heap-cloud. It has a flat base and rises in billowy rolls straight upward for perhaps twenty thousand feet. It is usually seen *Heap-clouds.* in the afternoon and is extremely brilliant in light, is tinged by various hues as the sun declines, and, again, has various colors in its shadows. There may be many of these heap-clouds cast in the form of a snowy mountain range, and when seen above the mountains themselves they produce the fairy illusion of mountains upon mountains, peaks upon peaks, rising upward into the blue. They seem to *Sky illusions.* lift and loom unendingly, and, though as unsubstantial as the baseless fabric of a dream, how beautiful they are to watch in their various forms, lights, and colors!

The air currents from the valley may rise still higher than the tops of the towering cumulus, and may carry a less percentage of moisture from having had some of it squeezed out by the chilling of the temperature on the way up. The remainder may reach thirty or forty thousand feet above the valley and then condense into the form of the stratus, or the long, flat cloud that sometimes stretches across the heavens. It is a frail, diaphanous cloud and takes many horizontal or flattened forms, but is usually not so marked or noticeable in light and color as the heap-cloud.

Still higher than this the air currents may rise in invisible form, having still less percentage of moisture in their composition, until they reach a zero temperature, perhaps sixty or eighty thousand feet in air. Then when they condense it is not into vapor but into ice spicules or ice-dust—vapor frozen into tiny particles that float in the same manner as hoarfrost. The forms these highest clouds, the cirri, take are varied, and because of their icy make-up they are very brilliant in color and light, very translucent, very diaphanous. Their appearance is never quite the same from hour to hour.

"At dawn a mist of sapphire faintly glowing,
    At noon a wave of white,

With curling edges into plumage growing
    And spreading wings of flight.

"At dusk a flame above the mountains gleaming,
    —The dead Day's funeral pyre—
A scarlet banner on the blue vault streaming,
    Or golden hair afire.

"At night a mesh of bright auroral veiling,
    Waving in broken bars,
A nebula across the sky depth trailing
    Tangled with silver stars."

Radiant and glowing when seen from the valley, the cirri are sometimes more piercing in light, more scintillant in color, when seen from the mountain's top. From either point of view they are the glory of the upper sky— beacon lights that seem to beckon on to unknown regions of space. *The glory of the upper sky.*

Quite different from the stratus or the cirrus is the nimbus—the true rain-cloud. It may be formed in the same way as the others, but from oversaturation may reach its precipitation-point when only a few thousand feet from the valley and send forth torrents of rain. Again, when a sirocco blows across the Mediterranean and is tilted up from the plains of Lombardy toward the snowy Alps, or when *The nimbus or rain-cloud.*

Chinook winds.

Falling rain in the valley.

Storm in the mountains.

a warm chinook wind, moisture-laden from the Pacific, creeps up through the passes of the Canadian Rockies, there is condensation into nimbus clouds all along the line of cold contact and a heavy precipitation of rain is not the unusual result. The clouds themselves are not brilliant in color or light and are usually seen as a huge, overspread veil of gray.

Rain in the mountains is not very different from rain elsewhere except that you usually see it against a background of mountain slope or standing timber. It falls straight and heavy, or, if there is wind, then waving in diagonal lines, and possibly filling the whole valley with a gray mist that cannot be penetrated by the eye. Sometimes it is furious and lashes crests and minarets and mountain walls with a force unseen elsewhere, pouring down in torrents and washing bare surfaces almost in waves. With it at these times comes lightning that seems to run in zigzags from cloud to cloud or peak to peak, and then again in rivulets that stream down into the valley with a great crash of accompanying thunder. The reverberation of the rocky walls, perhaps, creates the impression of there being more lightning than really exists. Thunder in mountain valleys or even among the high peaks is usually terrific in its sound.

In times of wide-spread storm the clouds hang down low in the valleys and are ragged on their under edges. Frequently parts of them get entangled in the tree-tops of the high timber and tear into fragments like thin bats of cotton, or they drift helplessly about bumping themselves to pieces against rocky escarpments. The appearance of these storm-clouds when seen from the upper peaks looking down upon their tops is, of course, entirely different. The mass of clouds is then brilliantly illuminated by the sun and in appearance is not unlike a heavy, sunlit fog rolling in from the sea. Its hue is a radiant white or silver and its movement is billowy or rolling.

After a storm has passed there is often a bright effect produced by the sun shining upon the rear of the passing clouds and converting the cloud-mounds into visions of dazzling whiteness. Below the white domes and turrets of cumulus there is occasionally a fall of rain wherein the rain-drops are struck by the sun and changed into falling diamonds. And, of course, there is the common but always beautiful arch of the rainbow to span the valley at some time during the storm. But perhaps any and all of these features are seen better from the valley than the peak. In fact, a storm looked down upon from the sunlit upper regions is a rather disappointing affair because you do not see the storm.

*The un-
moved
peaks.*

The peaks themselves seem to regard storms very lightly. They are unmoved and unchanged. The roar and the thunder, with the swish of rain and the whirl of winds, are only so much tempest in a teapot to them. A cloudburst hurled against the Matterhorn or a blizzard roaring about Mount Stephen is pitiful in its impotence. After many centuries of wear the storms may drag down the granite needles and lay them in the dust, but the wear of a single storm, no matter how severe, is as nothing. The walls shake not, the underbasing does not tremble, the foot-hills are not moved. As for the snow-crowned peaks, they lift their heads serenely into the sunlight and give no intimation of the attack.

*Mountain
serenity.*

The splendid repose of the upper peaks is half of their sublimity and is productive of perhaps half of our admiration. By contrast with the noisy brawling of the winds and waters they appear so restful, so majestic, so beyond all change or sway or influence. Many causes, perhaps, lead up to our regarding the mountains as sublime, as we shall see hereafter, but not the least of them is mountain serenity.

CHAPTER XII

THE RANGES

THE white peaks of the great ranges seem the lonely survivors of the Great Ice Age. When the ice-sheet slipped away from the equator, retreating on either side toward the poles, the white caps of the high mountains were left behind—the only remnants of the snow cover left in the temperate belt. White they remain to us at the present time and white they will probably continue for centuries to come. Their altitude and present temperature make possible their cloaks of

perpetual snow. The snow mantle is always losing at the bottom—running away in brook and river—but it is always being renewed at the top by new snowfalls and glacier formations. Hence, it is continuous in kind if not everlasting in the part.

High up on the plateaus of the world the white peaks stand clustered in masses like the Alps, or lifted into an enormous wall like the Himalayas, or strung out in a long, wind-

216

ing line like the Andes and the Rockies. The Alps are more localized in area than the others and perhaps for that reason more violent in folding and splintering. Enormous and imposing as they appear, they are outbulked by the Himalayas—the "High Asia," the true roof of the world. These come across India into the Caucasus and almost touch shoulders with the Alps; they extend down into the Malay Peninsula and, with interruptions, run into the islands of the sea. This immense wall of rock is what the Arabs call "the stony girdle of the world," and it is well named. The Andean-Rockies are more regular and perhaps less imposing in bulk, but they run from Cape Horn to the Behring Sea—practically from pole to pole—in an unbroken sequence and form a chain of white half-way around the globe.

The northern Rockies in places have the regularity of sea waves, and a lively imagination might regard them as drifted into their present place by some furious storm in the dark, abysmal time ere the stars sang together—a storm so furious that the surface of the earth was lifted and rolled into waves as though it were water.

"You can see the upward bend
    With the gathering of forces,
And the wave-crests in their trend
    Breaking north in foaming courses.
Were the mighty winds set blowing
    So terrific in their mood
That they started mountains flowing
    And then left them where they stood?

"Lo! the snowy tops still stand
    Petrified in waving motion,
Tossing whitecaps of the land
    Rising from a rocky ocean.
They are lifting, they are waving
    Frozen arms against the sky,
Crested spectres once set raving
    In a storm age long gone by."

Placing of the ranges.

But the placing of the mountain ranges, the lines they follow, the continents they traverse are not supposed to have quite so fantastic an origin as that. Dana long ago discovered that the greatest mountain ranges stand opposite the greatest ocean basins—the lower portion of South America which is opposite eastern Asia being almost the only exception.

Elevation of continents.

The elevation of one portion of the globe is supposed to have something to do with the depression of another portion of it. The continents themselves, if we choose so to regard them, are merely mountains, uplifts of the land above the sea. Were they mere accidents or was there method in their making? And the ranges heaved up along their back—the long ranges like that from Alaska to Tierra del Fuego—are they merely casual wavings of an oversqueezed crust?

Regarding such questions, science hesitates or speculates, while philosophy and religion go by on the other side. The problem is as yet too much for the mental grasp of man. Imagination pictures a primitive world, whirling in space as a shining ball evenly covered with water, and sees it, by coolings and changings, sinking in certain spots, and in those spots the gradual settling of the seas. The waters would thus be drawn off from the parts of the globe that did not sink, and, by comparison, they would become the elevated portions of the dry land, or the continents. That much might be believed about the general distribution of land and water and still give no hint about our mountain ranges. The long line of the Andean-Rockies may have been caused by a lateral thrust, along a line of weakness in the crust; but how did the pressure happen to come there by the edge of the sea and extend half-way around the globe from north to south? In spite of scientific achievement there still remains some ignorance about the great ranges.

Ignorance of the ranges.

The ignorance is almost as marked topographically as geologically. All of the ranges have, of course, been mapped and measured and duly described. The polar regions form an exception, though we are given hints of vast mountains at the south and large continental tracts lying to the north. But, aside from the scientific and governmental charting of the ranges, there is no wide diffusion of knowledge about them or experience in or with them.

The Alps the best known.

The Alps are the best known of all the mountains, thanks to their accessibility, but how few there are who know about the Rockies and how fewer still about the Andes! The Himalayas and the Caucasus are in lands long known and still thickly populated, but the great ridges remain unexplored and the people see them only from afar off. An Alpine climber

The un-known Hima-layas.

breaks into them occasionally, as he has done into the Andes, but the natives regard him as something of a madman, and the outside world puts him in a class with the arctic explorer. The world at large has not as yet made any appreciable impression upon the mountains. The great preserves of the high ranges are still practically unknown, untouched.

Nor have the people of the past ever had a proper appreciation of the mountains, possibly owing to a like ignorance of them. Mountain-climbing for centuries was a necessity, perhaps, of goatherds, but never a pleasure of the learned or the æsthetic or the merely curious. In 1336 Petrarch made the ascent of Mount Ventoux (6,200 feet), in France. He lived near the mountain and yet had the ascent in mind for years before he undertook it, so desperate and wild an adventure did it seem to him. In a description of it, written to the monk Dionysius at the University of Paris, he groans over the ascent like a modern alpenstock tourist. When he got to the top he looked for "the view," and saw merely the Bay of Marseilles. He wondered that he could not see the Pyrenees! Then he sat down, pulled out a copy of Saint Augustine's *Confessions*, and began moralizing on the grandeur of the human soul and the consequent meanness of mountains.

Just such an exalted notion of the ego and

*Apprecia-
tion of
mountains.*

*Petrarch
climbing
Mount
Ventoux.*

contemptuous opinion of the cosmos obtained with the polite and learned down to a very recent date. The mountain ranges were regarded as either a nuisance or a terror. They were almost the last of earth to win love and admiration. Leslie Stephen quotes Bishop Burnet as referring to them as "vast undigested heaps of stone," and again as exclaiming: "Look upon these great ranges, in what confusion do they lie; they have neither form nor beauty, neither shape nor order, no more than the clouds in the air. Then how barren, how desolate, how naked are they! How they stand neglected by nature! Neither the rains can soften them nor the dews from heaven can make them fruitful." Probably most of the people of the time shot as wide of the mark as that.

Bishop Burnet on mountains.

Mountain poetry.

Then, after the hopelessly unsympathetic, came the bombastic, with poets like "Seasons" Thomson, who used the mountains as so much sonorous copy. There was little more understanding or sympathy than before. Nature was not a thing in itself but a state of mind or a mood. It was so even with the romanticists. One by one the inspired ones had their try at the mountains, but which one of them reached to Alpine heights? Byron dragged down Mont Blanc by likening it to a monarch seated upon a throne, crowned with a diadem of snow, and Coleridge turned the same mountain into a mole-hill by describing it as piercing the "ebon mass" of air around it as "with a wedge"! Shelley is just as inadequate because overdone.

Byron on Mont Blanc.

> "The keen sky-cleaving mountains
> From icy spires of sunlike radiance fling
> The dawn, as lifted Ocean's dazzling spray,
> From some Atlantic islet spattered up,
> Spangles the wind with lamplike water-drops."

The glitter of it is more like the prisms of an old-fashioned chandelier than the frozen spires of the peaks. Once more hear the great Goethe describe a snowfall on the Alps:

Goethe on the Alps.

> "Yesterday brown was still thy head, as the locks
>   of thy loved one
>   Whose sweet image so dear silently beckons afar.
>   Silver-gray is the early snow to-day on thy
>   summit,
>   Through the tempestuous night streaming past
>   over thy brow."

Wordsworth never wrote anything tamer nor "Ossian" Macpherson anything more empty.

The real poetic allusions to the mountains that have strength in them are in the Bible and Shakespeare. Only a very few minds seem to have risen to the mountains, and almost always it will be noted that those few refer to the mountains in simile or allusion rather than in description.

The artists of the paint-brush have fared little better than those of the pen, and in the main have followed in the same strain as the writers. The exceptions are men like Dürer and Altdorfer, in Germany, who knew the drawing of certain mountain forms exceedingly well; and Titian at Venice, who grew up in the Cadore Mountains and never forgot them. Titian's mountains in the background of such pictures as the "Presentation," in the Venice Academy, are superb, and several of his followers (Palma and Bonifazio) show, perhaps, a transmitted knowledge of mountain arrangement. Later on came the egotists, who thought of nature as merely a second fiddle in the orchestra—something that would, perhaps, help out human expression. Salvator Rosa used precipices and storm-swept peaks merely to suggest his stormy mood; Claude and Poussin used them as a classic background;

Delacroix employed them to help out a romantic story. It was all of it more or less distortion though decorative and appropriate enough as a filling up of the canvas.

Courbet was more truthful and far better, especially when he did portions only of ranges or precipices or rocky retreats. Turner sketched among the Alps, and Segantini lived and painted among them during the latter part of his life; and from these two came many suggestions of mountain beauty. But they, in common with all the others, had fundamental difficulties to deal with. Not all the eloquence of Ruskin, with the indorsement of the later Hamerton, could entirely remove those difficulties. Ruskin insisted that the color of the Alps was the most lovely in all the world; but then, his world extended no farther than a limited portion of western Europe. He admired Alpine coloring in spots rather than in masses. The small patches of bell-gentian, oxalis, and violets seemed to take his fancy more than the gray peaks or the blue sky, but he had a very true word for the mountain blues, roses, and purples. He had, indeed, such a sincere love for mountain beauty that it would not let him see any beauty whatever in the flat lands. His analysis of the Alps was, however, a keen exposition notwithstanding its fantasies. All that he had to say about structure, about peaks and precipices and snows and waterfalls, was quite true. And he had the added virtue of literary style which commanded the attention of the art world.

Painters could not follow Ruskin.

But somehow the painters never could follow Ruskin in the mountains with any success. He thought the mountains elevated and ennobled mankind, as Buckle, *per contra*, thought they dulled the intellect; but the painter people cared nothing about such abstractions. They could not see the color or get the atmospheric setting or compress the pattern without dwarfing or belittling the theme. Some there were who insisted the Ruskin colors did not exist; others there were who found the purples, blues, and dark greens, but found them too cold. And completely over-

Difficulties of mountain-painting.

awed by the vastness of form. A drawing of a mountain range by masses of color was possible but when done was not probable. A linear drawing with light and shade was even worse, for it became rambling and panoramic. Yet there is no doubt whatever about the beauty of the mountains in themselves.

Why cannot they be reduced to canvas? Is it impossible to poetize or visualize them? Are they beyond the pale?

To say that the high mountains, though sublime in their form and color, are not pictorial is to assume the whole question; yet, nevertheless, that is the proper answer. The qualities of sublimity in the mountains, such as bulk and mass, are the very qualities that cannot be realized or placed on canvas. The abrupt lift, the height, the perpendicular line that counts in the mountain for majesty and dignity—just as the horizon-line in the sea makes for repose—are things that cannot be adequately suggested in their enormous size. The necessary reduction in scale loses tremendously in suggestiveness. Moreover, there is still another quality that comes with the mighty uplift—the quality of looming in the peak. At times the whole peak seems to ride the blue sky and comes looming forward and upward above us. Such illusions or impressions are produced only by the actual bulk and lift, and once more they cannot be reduced to canvas without becoming dwarfed, tame, and expressionless.

In a narrower sense the pictorial—that

Mountains not pictorial.

Illusions and impressions.

is, something fitted for a picture—means primarily a decorative arrangement of form and color, an agreeable pattern that can be scaled to a certain-sized canvas. The mountains hardly lend themselves to the pattern, and their placing upon canvas usually results

in some crowding and mutilation. If seen close to view the huge forms shut out both sky and foreground; if seen from a lowland distance in many peaks and groups the picture inclines to the spectacular, the grandiose, the topographic. The view from near Turin looking northward to the Alps is magnificent; so, too, is the panorama of the snowy sierra from Granada or the Mexican mountains from Zacatecas; but none of these scenes would shape up or cut up into a picture. They are

too mappy. And, again, the jagged peaks would result in an irritating saw-edge for a sky-line—a series of perpendicular and diagonal lines without a round or flowing line to break the monotony. The pattern would not be agreeable in form.

When the matter of the color in the pattern is considered the high mountains will once more be found lacking. The color is too cold. Whites and blues and purples may be true, appropriate, and very acceptable in nature, and yet be forbidding in art. They have not pictorial charm or loveliness; they do not win or enchant us on the canvas. Harmony with the great colorists has usually meant a balance of warm tones by cold ones, but the mountains have no warm tones except in unimportant spots. Even the greens of the timber belt are cold greens that border upon purple, and as for the reds and yellows, they do not exist except, again, sporadically. On the canvas the snowy peaks would be given in three or four large planes of color. In the foreground would, perhaps, come a strip of green, then a higher strip of gray-green representing the uplands, then a zigzag of white showing the peaks, and lastly, at the top of the canvas, a flat band of blue sky. Would that make up an attractive color pattern? Not all the skill of the Courbets and the Rousseaus, nor all the truth of the latter-day Scandinavian impressionists, has been able to make it so.

The coldness of the color.

Absence of warm hues.

When one thinks of light and atmosphere in mountain pictures the difficulties simply increase. The light has little or no warmth and is too clear, too penetrating. It robs the

Light and atmosphere.

scene of all mystery and is inclined to be glaring. The shadows are often supremely lovely, but so delicate that when accurately painted they fail in carrying-power. The air, again, is too thin, too intangible, for a blending and holding medium. At great heights it practically goes out, and any accurate rendering of it would give the impression of a vacuum in the upper space. It is not dense enough to create what is called the "envelope" of a picture.

Moreover, painting is of a middle quality between a thought and a thing—"the union of that which is nature with that which is exclusively human." The human element in thought, feeling, or emotion should be in it, otherwise it is no better than so much mechanical tracery. But when nature appears upon so stupendous a scale, what chance is there for human personality to come in or for emotion or feeling to make itself manifest except by distortion? How could the most accomplished emotionalist give you his feeling about the Alps or the most uncompromising impressionist his impression about the Himalayas without sacrificing objective truth? Once more the forms are too colossal.

Of course, there have been some good pictures painted of the mountains, in spite of objections that may be raised; but they have usually succeeded by compromising or evading difficulties. When the high peaks and the snow are eliminated an arrangement of green, gray, and blue is left that is certainly less difficult to harmonize than the entire scene. Still, even the green mountains have not been too successfully handled in art. Claude and Poussin painted them a tapestry green, such Dutchmen as Ruysdael gave them a pincushion olive-green, and the Englishmen, Gainsborough and Crome, toned them to the key of brown leather; but it was all a rather sorry distortion of mountain color. Painters in America—A. H. Wyant and Homer Martin for instances—have come nearer to the truth of both nature and art by using warm light and a richly colored foliage. They, however, have dealt more with low Adirondack hills than high mountains. *Painting green mountains.* *Claude, Ruysdael, Gainsborough.* *Wyant and Martin.*

The paintable mountains—the ones that would make up the decorative pattern in form and color, light and air—are those of the desert. The ranges of Arizona, for example, are not marked by whites, greens, and cold *The desert ranges paintable.*

purples. The slopes are bare of trees and only a gray yellow of chaparral or cactus or bunch-grass appears here and there in spots. The red of porphyry shines through the thin covering and gives the slopes and peaks a ruddy hue, the air, especially in the heat of summer, is opalescent or, at times, lilac-hued, and the light is often pale orange or warm pink. Seen through such media the huge foldings and breakings of porphyry and granite float weirdly in a sea of color, look mysteriously splendid. At sunset it is not only the sky and air that are deluged with color, but the bare peaks take fire from the western light and glow like turrets of molten metal.

Here is not only warmth of color but an atmosphere that is thick enough to blur form and keep it from too great prominence. Here also is a light sufficiently diffused to do away with all sharp contrasts. The conditions for a decorative pattern on canvas are existent in the desert landscape, but the painter seems not yet to have arrived. Decamps in Turkey and Fromentin in Algiers were too much interested in humanity to care much about their mountain backgrounds, and here in America, though there have been some excellent attempts, the vastness of the ranges and the wilderness of color have bothered such of our landscapists as have made the trial. The mountains of the desert really call for a painter with a different color sense from any of his predecessors. His palette must be limitless and his sense of harmony all-embracing. That is merely another way of returning upon ourselves and declaring the difficulties of the mountain on canvas. Evidently the high peaks were not intended for exploitation in art.

A new field.

Is it matter of regret that the painter cannot lash the Himalayas to the mast or spread the Andes on canvas? On the contrary, there is, perhaps, a satisfaction in thinking that some things in nature are beyond man's power. It is a pleasure to think of the mountains, for instance, that no one can till them or fell them or destroy them, that commercially or economically they cannot be capitalized and sold on the market, that even artistically they are hardly to be captured and lugged into the drawing-room. Evidently, the best that man can do about them is to wonder over them and admire them. And what things to wonder over are the great ranges! To the north of

Things beyond man's reach.

The unattainable ranges.

Italy lifts the snowy boundary of the Alps, to the north of Persia stretches the grim Caucasus, to the north of India rises the giant barrier of the Himalayas. Merely exposed portions of the crust, but how very astonishing they are in mass and volume! Shall we ever cease to exclaim over them? Shall we ever become so familiar with them that they will appear commonplace?

When everything is known.

When all the world is explored, charted, diagrammed, when everything is known in and about its surface, what a very dull place it will be to live in! All the poetry and glamour of it will have vanished. We are fortunate in our age that there is still some ignorance left, still something to marvel over. The great

The wonder of the unknown.

mountain ranges have not ceased to be a source of mystery. Again and again, as we ride away, we turn in the saddle to look at their massive forms against the sky. They keep drawing us with a new look or an old lure. They are not paintable, they are not habitable, they are not wholly understandable, but, perhaps for that very reason, they are wonderful. May they always remain so!